WITHDRAWN
UTSA LIBRARIES

BURT FRANKLIN: RESEARCH & SOURCE WORKS SERIES
Selected Studies in History, Economics, & Social Science:
n.s. 14 (a) Classical & Oriental Studies

ROMAN CRAFTSMEN AND TRADESMEN OF THE EARLY EMPIRE

ROMAN CRAFTSMEN AND TRADESMEN OF THE EARLY EMPIRE

BY

ETHEL HAMPSON BREWSTER

**BURT FRANKLIN
NEW YORK**

Published by LENOX HILL Pub. & Dist. Co. (Burt Franklin)
235 East 44th St., New York, N.Y. 10017
Reprinted: 1972
Printed in the U.S.A.

Burt Franklin: Research and Source Works Series
Selected Studies in History, Economics, & Social Science:
   n.s. 14 (a) Classical & Oriental Studies

Reprinted from the original edition in the Princeton University
   Library.

Library of Congress Cataloging in Publication Data

Brewster, Ethel Hampson, 1886-1947.
  Roman craftsmen and tradesmen of the early empire.

    Reprint of the 1917 ed.
    Originally presented as the author's thesis, University of Pennsylvania,
1915.
    Bibliography: p.
     1. Industrial arts—History. 2. Rome—Antiquities. 3. Latin poetry—
History and criticism. I. Title.
DG107.B8 1972   338'.0937'6   72-81956
ISBN 0-8337-4822-X

PREFACE

Interest in the subject of this dissertation was stimulated by a study of the private life of the Romans. Investigations in the satiric writers, whose exaggerations form the usual lens through which the common view of Roman life is gained, left the impression that the picture which is habitually displayed gives no accurate portrayal of ordinary society, but a distorted glimpse of court life, high society, and the social struggle therein. The thought presented itself, therefore, that it would be possible to argue a human existence for even "the butcher, the baker, the candlestickmaker."

The writer is glad to take this opportunity to acknowledge with gratitude her indebtedness to the University of Pennsylvania for the privilege of holding for a year and a half Bennett Fellowships in Classics. She desires also to express her earnest appreciation to Professor John C. Rolfe, Professor Walton B. McDaniel, Professor Roland G. Kent, and Assistant Professor George D. Hadzsits, whose helpful criticisms and suggestions have been of the greatest practical assistance, while their scholarly attainments have proved a never failing source of inspiration and encouragement.

E. H. B.

CONTENTS

Verbaque provisam rem non invita sequentur
Hor. *Epist.* 2. 3. *311*

		Page
BIBLIOGRAPHY		vii
INTRODUCTION		xi
I.	Aerarii Ferrarii	1
II.	Argentarii III. Aurifices IV. Caelatores	6
V.	Caupones	9
VI.	Centonarii	13
VII.	Cerdones	13
VIII.	Coci	13
IX.	Coriarii	18
X.	Dendrophori	19
XI.	Fabri	19
XII.	Ferrarii	19
XIII.	Figuli	19
XIV.	Fullones	20
XV.	Institores	22
XVI.	Lanii	27
XVII.	Mangones	29
XVIII.	Mercatores XIX. Negotiatores	30
XX.	Pistores	40
XXI.	Praecones	44
XXII.	Sutores Cerdones	53
XXIII.	Tabernarii	60
XXIV.	Textores	74
XXV.	Tignarii	77
	Collegia Fabrum Centonariorum Dendrophorum	79
XXVI.	Tonsores	87
CONCLUSION		94

BIBLIOGRAPHY

Tempus non tam stilo quam inquisitioni instituti operis prope infiniti et legendis auctoribus, qui sunt innumerabiles, datum est. Quint. *Praef.*

All the important commentaries on the satiric writers of the early Empire have been examined for this dissertation. The text editions usually followed are:

Q. Horati Flacci Opera, Wickham-Garrod. 2d ed., Oxford, 1912.
A. Persii Flacci, D. Iunii Iuvenalis, Sulpiciae Saturae, Jahn-Bücheler-Leo. 4th ed., Berlin, 1910.
Petronii Saturae et Liber Priapeorum, Bücheler-Heraeus. 5th ed., Berlin, 1912.
M. Val. Martialis Epigrammta, W. M. Lindsay. Oxford, 1902.

Merely a selected bibliography is given here, since a complete list, especially of books of a general nature, would include all works on Roman private life. For an exhaustive bibliography on Roman Industrial Corporations, see Waltzing 1. 6–30.

For books mentioned below, except when it is necessary to distinguish between several works of one writer, only the author's name, with the number of the volume and the page, is cited in the footnotes. For Latin sources, the abbreviations of the *Thesaurus Linguae Latinae* are used, with a few common exceptions like CIL., Hor. *Sat.* Cross references to this dissertation are designated as follows: See p.—, n.—.

Abbott, F. F., Society and Politics in Ancient Rome. New York, 1909.
    The Common People of Ancient Rome. New York, 1911.
Beck, C. H., The Age of Petronius. Cambridge (Mass.), 1856.
Becker, W. A., Gallus. Trans. by F. Metcalfe. London, 1895.
Biot, E. C., De l'abolition de l'esclavage ancien en occident. Paris, 1840.
Blair, W., An Inquiry into the State of Slavery amongst the Romans. Edinburgh, 1833.
Blümmer, H., Die gewerbliche Thätigkeit der Völker des klassischen Altertums. Leipzig, 1869.
    Technologie und Terminologie der Gewerbe und Künste bei Griechen und Römern. 5 vols., Leipzig, 1875-1887; vol. 1, new ed. 1912.
    Die römischen Privataltertümer. Handbuch der klassischen Altertumswissenschaft, von I. Müller, vol. 4. 2. 2, Munich, 1911.
Bürger, R., Der antike Roman vor Petronius. Hermes, vol. 24 (1892), pp. 345-358.
Carter, J. B., Religion of Numa. New York, 1906.
Cauer, F., Die Stellung der arbeitenden Klassen in Hellas und Rom. Neue Jahrbücher für das klassische Altertum, vol. 3 (1899), pp. 686-702.
Cooper, L., A Concordance to the Works of Horace. Washington, 1916.
Cunningham, W., Western Civilization in its Economic Aspects. Vol. 1, Ancient Times. New York, 1902.
Daremberg et Saglio, Dictionnaire des antiquités grecques et romaines. Paris, 1881—
Davis, W. S., The Influence of Wealth in Imperial Rome. New York, 1910.

BIBLIOGRAPHY

Dessau, H., Inscriptiones Latinae Selectae. Vol. 2.2, Berlin, 1906.
Dill, S., Roman Society from Nero to Marcus Aurelius. New York and London, 1905.
Donaldson, J., Woman: Her Position and Influence in Ancient Greece and Rome. London, 1907.
Drumann, W., Arbeiter und Communisten in Griechenland und Rom. Königsberg, 1860.
Duruy, V., History of Rome. Vols. V and VI, The Empire and Roman Society. Trans. by M. M. Ripley and W. J. Clarke. Boston, 1883.
Ernesti, J. A., De Negotiatoribus Romanis. Opuscula Philologica Critica. Leyden, 1776.
Fowler, W. W., Social Life at Rome in the Age of Cicero. New York, 1909.
Frank, T., Mercantilism and Rome's Foreign Policy. Amer. Hist. Review, vol. 18 (1912-1913), pp. 233-252; also incorporated in:
Roman Imperialism, ch. 14, pp. 277-297: Commercialism and Expansion. New York, 1914.
Friedländer, L., Roman Life and Manners under the Early Empire. Trans. of 7th ed.: vol. 1 by L. A. Magnus, vols. 2 and 3 by J. H. Freese, vol. 4 by A. B. Gough. New York, 1909-1913.
Giese, P., De personis a Martiale commemoratis. Greifswald, 1872.
Gilbert, O., Geschichte und Topographie der Stadt Rom im Altertum. 3 vols., Leipzig, 1883-1890.
Godefroy, J., Codex Theodosianus cum perpetuis commentariis. New ed. by J. D. Ritter, Leipzig, 1743.
Gusman, P., Pompeii, the City, its Life and Art. Trans. by F. Simmonds and M. Jourdain. London, 1900.
Harcum, C. G., Roman Cooks. Johns Hopkins Dissertation. Baltimore, 1914.
Cp. Rankin, E. M., The Role of the Μάγειροι in the Life of the Ancient Greeks. Chicago, 1907.
Hirschfeld, O., Der praefectus vigilum in Nemausus und die Feuerwehr in den röm. Landstätten. Sitzungsber. der Wiener Akad., vol. 107 (1884), pp. 239-257.
Inge, W. R., Society in Rome under the Caesars. New York, 1901.
Jahn, O., Darstellungen des Handwerks und Handelsverkehrs auf Vasenbildern; Darstellungen antiker Reliefs, welche sich auf Handwerk und Handelsverkehr beziehen. Leipzig, 1861.
Johnston, H. W., The Private Life of the Romans. Chicago, 1907.
Jordan, H., Topographie der Stadt Rom im Altertum. 2 vols., Berlin, 1871-1907.
Knapp, C., Business Life as Seen in Horace. P.A.P.A., vol. 29, pp. xliv-xlvi.
Kornemann, E., Collegium. Real-Encycl., vol. 4. 1, pp. 380-480.
Fabri. *Op. cit.*, vol. 6.2, pp. 1888-1925.
Kühn, G. B., De opificum Romanorum condicione privata quaestiones. Halle, 1910.
Liebenam, W., Zur Geschichte und Organisation des römischen Vereinswesens. Leipzig, 1890.
Mackenzie, W. M., Pompeii. London, 1910.
Marquardt, J., Das Privatleben der Römer. 2 vols., Leipzig, 1886.
Mau, A., Pompeii, its Life and Art. Trans. by F. W. Kelsey. New York, 1904.
Maué, H. C., Der Praefectus Fabrum. Niemeyer, 1887.

Miller, A. B., Roman Etiquette of the Late Republic. Univers. of Penn. Dissertation, 1914.
Mommsen, T., De collegiis et sodaliciis Romanorum. Kiel, 1843.
History of Rome. Trans. by W. P. Dickson. 5 vols., New York, 1911.
Nicolson, F. W., Greek and Roman Barbers. Harv. Stud., vol. 2, pp. 41-56.
Overbeck, J., Pompeii. 4th ed., Leipzig, 1884.
Pauly-Wissowa, Real-Encyclopädie der classischen Altertumswissenschaft. Stuttgart, 1894—
Peck, T., The Argiletum and the Roman Book Trade. Class. Phil., vol. 9 (1914), pp. 77 f.
Peet, T. E., The Stone and Bronze Ages in Italy and Sicily. Oxford, 1909.
Pellison, M.. Roman Life in Pliny's Time. Trans. by M. Wilkinson. New York, 1897.
Platner, S. B., The Topography and Monuments of Ancient Rome. 2d ed., Boston, 1911.
René, M., La vie privée des anciens. Vol. 3, Le travail dans l'antiquité. Paris, 1882.
Richter, W., Handel und Verkehr der wichtigsten Völker des Mittelmeers. Leipzig, 1886.
Sandys, J. E., A Companion to Latin Studies. Cambridge, 1910.
Saulnier, A. le, Du travail salarié à Rome. Paris, 1888.
Schanz, M., Geschichte der Römischen Litteratur. Handbuch von I. Müller, vol. 8.2.2, Munich, 1913.
Schreiber, T., Atlas of Classical Antiquities. New York, 1895.
Smith, W., Dictionary of Greek and Roman Biography and Mythology. 3 vols., London, 1844-1849.
Dictionary of Greek and Roman Antiquities. 2 vols., 3d ed., London, 1901.
Speck, E., Handelsgeschichte des Altertums. Vol. 3, Leipzig, 1906.
Teuffel-Schwabe, History of Roman Literature. Trans. of 5th ed. by G. C. W. Warr. London, 1891-1892.
Thomas, E., Roman Life under the Caesars. Trans. New York, 1899.
Tucker, T. G., Life in the Roman World of Nero and Saint Paul. New York and London, 1910.
Typaldo-Bassia, A., Des classes ouvrières à Rome. Paris, 1892.
Walde, A., Lateinisches etymologisches Wörterbuch. Heidelberg, 1910.
Wallon, H. A., Histoire de l'esclavage dans l'antiquité. 3 vols., Paris, 1847.
Walters, H. B., History of Ancient Pottery. New York, 1905.
Waltzing, J. P., Étude historique sur les corporations professionnelles chez les Romains. 4 vols., Louvain, 1895-1900.
West, L. C., The Cost of Living in Roman Egypt. Class. Phil., vol. 11 (1916), pp. 293-314.
Wezel, E., De opificio opificibusque apud veteres Romanos. Berlin, 1881.
Wissowa, G., Religion und Kultus der Römer. Handbuch von I. Müller, vol. 5.4, Munich, 1912.

INTRODUCTION

In the various volumes that have been written on "The Private Life of the Romans," the account of their industrial population is usually introduced by almost stereotyped expressions, stating that "Unfortunately our information concerning all this class is very inadequate. The Roman writers—historians, philosophers, rhetoricians, and poets—have extremely little to say about the humble persons who apparently did nothing to make history or thought. They are mentioned but incidentally, and generally without interest, if not with some contempt."[1] Those, therefore, who have made a special study of the condition of "these small people"[1] have sought more fertile fields than that of literature proper. A. Typaldo-Bassia, for instance, in an excellent treatise entitled *Des classes ouvrières à Rome*, takes the Pandects as his chief source for a detailed study of the legal status of Romans who worked for their living. Again, G. B. Kühn's dissertation, *De opificum Romanorum condicione privata quaestiones*, is an invaluable storehouse of inscriptional references to Roman craftsmen; these have been collected with laborious care from the *Corpus Inscriptionum Latinarum*, and are arranged in systematic lists classified according to occupations, with subdivisions for *ingenui*, *liberti*, and *servi*.

It would seem almost inexcusable then for a novice to step in where the more enlightened have hesitated to tread for want of a firmer foothold. But after all, a collection of civil laws that were codified by Justinian, who was Emperor of the East in the middle of the sixth century, is scarcely a satisfactory basis upon which to establish an investigation into the social conditions of five hundred years earlier. Though the Digest contains excerpts from Q. Mucius Scaevola, consul in 95 B. C., three-fifths of it is selected from Ulpian and Paulus,[2] whose "floruits" were in the reign of Septimius Severus, and much of it is due to still later jurists.[3] Now by the time of Septimius Severus, industrial corporations were coming more and more into the power of the emperors;

---

[1] Tucker 246 f.; the same sentiment, *passim*; practically the identical words, Fowler 26: "Unfortunately . . . we know little of its (Rome's) industrial population. The upper classes, including all writers of memoirs and history, were not interested in them." *Id*. 43: "These small people . . . did not interest their educated fellow-citizens, and for this reason we hear hardly anything of them in the literature of the time. Not only a want of philanthropic feeling in their betters, but an inherited contempt for all small industry and retail dealing, has helped to hide them away from us."

[2] Teuffel-Schwabe-Warr 2.§§ 376 f., 488.

[3] J. J. Robinson, *Selections from Roman Law*, Introd. 26 f. (New York, 1905).

and under Alexander Severus, in the cities, men of every occupation were compelled to form *collegia*, to which they were confined by such severe and restricting regulations that they completely lost all personal freedom and were reduced practically to the condition of serfs.[4] So in the country, *coloni* became bound to the soil.[5] Such a radical shift in the status of the working man not only affected the laws of the time, but must have influenced the selection of excerpts from earlier codifications and perhaps have instituted changes in them. This must be born in mind, even while it may be agreed that legal forms possess a peculiar tenacity, and that fairly accurate inferences for a given period may often be drawn from much later legislation. Then too, paradoxical as it seems, the letter of the law and the spirit of the age may be at variance, as the necessity for certain legislative *re*enforcemements and *re*vivals implies; and among the Romans, "customary law, growing out of the life and experiences of the community," was ever strong, continuing "to have validity as subsidiary law when not expressly abrogated by statute."[6]

As for epigraphical sources, although inscriptions have proved of inestimable worth in filling lacunae along all lines of antiquarian research and in corroborating the theories of savants, yet with but chance exceptions they are as cold and inflexible as the stone or metal on which they are incised.

For vigorous, animated, actual life, therefore, whether of the rich or poor, it seems necessary to turn to literature, and especially to the works of the satirist, "who must have a wide and comprehensive knowledge of his fellow men, . . . must be able to paint society in all its myriad hues."[7] Such a course is especially appropriate in an investigation into Roman conditions; for, to quote Quintilian, "Satira quidem tota nostra est;"[8] and as R. Y. Tyrrell has noted in his *Lectures on Latin Poetry*: "This was the way which Rome chose in which to 'hold the mirror up to Nature, to show Virtue her own feature, Scorn her own image, and the very age and body of the time his form and pressure.' "[9]

[4] Waltzing 2. 45 ff., 253 f.
[5] Cauer 699.
[6] Robinson, *op. cit.*, 11 f.
[7] Butler, *Post-Augustan Poetry*, 87 (Oxford, 1909).
[8] Quint. 10.1.93. Garrod in *The Oxford Book of Latin Verse*, Introd. XXII (Oxford, 1912), adversely criticises Quintilian's statement; but see Nettleship, *The Roman Satura*, 17-19 (Oxford, 1878), and Webb, *On the Origin of Roman Satire*, Class. Phil. 7 (1912). 188.
[9] Tyrrell 219 (New York, 1895).

Horace possessed such keen insight into human nature that to the admiration of Persius, he could keep a man his friend and evoke a laugh from him even while he made sport of him:

> Omne vafer vitium ridenti Flaccus amico
> tangit et admissus circum praecordia ludit,
> callidus excusso populum suspendere naso.[10]

Juvenal names "Everything Pertaining to Man" as the subject of his medley:

> Quidquid agunt homines, votum timor ira voluptas
> gaudia discursus, nostri farrago libelli est.[11]

And Martial's epigrams are, on his own assertion, merely "Little Stories of Real Life," with every page savoring of humanity:

> Hoc lege, quod possit dicere vita "Meum est."
> non hic Centauros, non Gorgonas Harpyiasque
> invenies: hominem pagina nostra sapit.[12]

Martial states also that it is especially characteristic of his Muse *dicere de vitiis*, (though he adds that it is his wont *parcere personis*);[13] and it is an enumeration of crimes and excesses that causes Juvenal to exclaim: "Difficile est saturam non scribere."[14] From an ethical point of view, therefore, it may not be "*un*fortunately" true, as suggested at the beginning, that the industrial population found little mention in the pages of the satirists. Moreover, whatever information is gleaned from these writers, either from between the lines or from stray allusions, may be more reliable than the sarcastic hyperbole and stinging rhetoric of their tirades on vice, in which, to press home their point, they are almost bound to exaggerate and magnify, while they keep their background and chance comparisons natural and normal.

It is the writer's aim, therefore, to discuss Roman craftsmen and tradesmen as depicted by the satiric writers of the early Empire. The expression, "satiric writers," is adopted in order to include not only Horace, Persius, and Juvenal, but also Petronius and Martial; the simpler word "satirists" is occasionally employed in the body of the dissertation with this broad signification. In lieu of a more comprehensive term, "craftsmen" is used to designate those whom the Romans called *opifices*. An attempt has been made to investigate in the authors above mentioned all passages relevant to our subject; to incorporate

[10] Pers. 1.116-118.
[11] Juv. 1.85 f.
[12] Mart. 10.4.8-10.
[13] *Id.* 10.33.10.
[14] Juv. 1.30.

the information secured into a connected account, with the aid of references from other sources by way of comparison or elucidation; and finally, to use this material as a basis for determining, so far as possible, the social status of Rome's industrial population during the period in question. Details have been added at times merely for the sake of interest, where they have seemed to vivify a character or enliven the narrative. There is, of course, much in the following pages that can lay no claim to novelty, but the work has all been done through independent research, and the result is submitted in the hope that there is a place for a monograph upon a subject which, in English at least, has received only scattered or fragmentary treatment.

ROMAN CRAFTSMEN AND TRADESMEN OF THE EARLY EMPIRE

> Quot capitum vivunt, totidem studiorum milia
>
> Hor. *Sat.* 2.1.27 f.

I

Aerarii   Ferrarii

Plutarch records a clever measure which, he says, was adopted by King Numa in order to unite the opposing Roman and Sabine factions by binding them together through common interests. The policy included the distribution of the people into eight *collegia*, consisting of flute-players, goldsmiths, carpenters and builders, dyers, workers in leather, tanners, braziers, and potters. All other craftsmen, the biographer continues, were collected into a single corporation.[1] According to Pliny the Elder, the third of these pioneer labor guilds to be established was composed of *fabri aerarii*.[2] Prolific excavations of actual articles, added to a wealth of literary material, give abundant evidence of the variety of bronze implements and utensils used by the Romans; it is readily inferred, therefore, that the ranks of the *aerarii* were numerous at all periods, especially since statuaries appear to have been classed with them.[3]

Practically nothing is to be gathered from Horace as to their social status, but two passages in which he refers to them demand special mention. In *Epist.* 2.1. 93-98 we read:

> Ut primum positis nugari Graecia bellis
> coepit et in vitium fortuna labier aequa,
> nunc athletarum studiis, nunc arsit equorum,
> marmoris aut eboris fabros aut aeris amavit,
> suspendit picta vultum mentemque tabella,
> nunc tibicinibus, nunc est gavisa tragoedis.

The expressions *nugari* and *in vitium labier* forbode ill; they are also misleading. Although Horace no doubt felt that war was the most serious occupation of man, yet the context shows that he is here really commending the Greeks for having ceased from war. He has been scoring the *laudator temporis acti*, and at this juncture he points presumably to the Athenians, who at the close of the Persian War had made

---

[1] Plut. *Numa* 17. See pp. 7, 13, 18, 20, 27, 53, 77.
[2] Plin. *Nat.* 34.1, *a rege Numa conlegio tertio aerarium fabrum instituto*. There seems to be no occasion for Waltzing's interpretation (1.63) that *aerarii* were third in rank in a hierarchy.
[3] Hor. *Epist.* 2.1.96; cp. 2.3.32 ff.; Mart. 9.68.6; Waltzing 1.52.

great advances in new forms of literature and art,[4] and had become the models of Horace's own time. His words can be no more disparaging to the *faber* than to *poeta*, whom he mentions in the same connection; and surely the soldier who threw down his arms at Philippi to become a littérateur[5] has little intention of criticizing devotees of the Muses.

The second reference is from those verses in the *Ars Poetica* in which Horace discusses harmony and proportion as the first requisites for unity in a work of art. By way of illustration, he mentions a sculptor who was successful in detail, but who failed with the *tout ensemble*:

> Aemilium circa ludum faber imus et unguis
> exprimet et mollis imitabitur aere capillos,
> infelix operis summa, quia ponere totum
> nesciet.[6]

The words of special interest to us, *faber imus*, form the subject of much discussion. Bentley, believing *imus* unintelligible, restored *unus*, which he claimed to be the reading of *codex Oxoniensis*; he suggested the translation "better than anyone else." This has been accepted by Orelli, Munro, Macleane, and others; but many editors, among them Krüger, Schütz, and Kiessling, still defend *imus*, which is found in most manuscripts; they interpret it as describing locality, or meaning "lowest in rank"; that is, "poorest," "most unskilful." The topographical explanation seems quite attractive in view of Horace's phrase in another Epistle, *Ianus summus ab imo*;[7] this may be supplemented by the *ad infimum Argiletum*[8] of Livy, and *circa imum Argiletum*[9] of Servius; in fact, Horace employs *imus* in the sense of "the end" in verses 126 and 152 of the *Ars Poetica* itself, note: "(Persona) servetur ad imum," "Primo ne medium, medio ne discrepet imum." Porphyrio's statement, too, is not to be disregarded; he maintains that the *ludus* of which mention is made was the gladiatorial school of Aemilius Lepidus, which by the commentator's day had been converted into the Baths of Polycleitus. He adds that Horace's words show that a *faber aerarius* had a shop *in angulo ludi*. Jordan argues that around the outer walls of the school there were probably shops which were rented by *fabri*; and that the last of these, facing the main street, was occupied by the *faber* of Horace's lines and by his successors, under the sign of the figure of Polycleitus;

[4] Aristot. *Pol.* 5.6.
[5] Hor. *Carm.* 2.7.9-14; *Epist.* 2.2.49-52.
[6] *Ib.* 2.3.32-35.
[7] *Ib.* 1.1.54.
[8] Liv. 1.19.2.
[9] Serv. on Virg. *Aen.* 7.607.

the business *signum*, then, gave rise to the name by which Porphyrio says the *ludus* was known after it had been turned into a Bath.[10] It is not characteristic of Horace, however, to label the subjects of his criticism so unmistakably as he would be doing in this case, if he is referring to *the* artificer in *the end* shop of a given district. It seems more plausible, therefore, to modify the above suggestions, and to concede that the satirist may be referring in an indefinite[11] way to *a* sculptor who might have been found at the lower end of the Ludus Aemilius where many copper-smiths had their *tabernae*.

In spite of the strength of the foregoing arguments, they are not completely satisfying. The fact is that to suit the context, and to preserve a proper balance with other illustrations, *faber* requires a qualifying adjective which means "unskilful." Horace's purpose in writing the *Ars Poetica* was to discourage from the pursuit of letters would-be writers who possessed no real literary ability: "Neither gods, nor men, nor bookshops," he warns in verses 372-374, "grant poets the boon of mediocrity." Throughout the Epistle, he emphasizes the importance of *ars* and *studium*. He criticizes Democritus (295 f.),

> Ingenium misera quia fortunatius arte
> credit;

and he lays down as his own dictum (409-411):

> Ego nec studium sine divite vena
> nec rude quid prosit video ingenium; alterius sic
> altera poscit opem res et coniurat amice.

Taking his precepts and examples at random, we note that it is a *callida . . . iunctura* (47 f.) that is praised in the use of diction; and *decor* (157), in the drawing of characters. The verses of Ennius are attacked, charged *ignoratae . . . artis crimine turpi* (262); and aspirants to literary fame are reminded of the athlete, who cannot be *indoctus* (380), if he is to be successful. In the face of this study of the Epistle as a whole, we are almost forced to believe that *imus* in verse 32 means "humblest,"[12] in the sense of "unskilful," and that it is but a synonym for *iners* (i. e., *in+ars*). This belief is confirmed, if we observe verse 31, which the

---

[10] Jordan, *Hermes* 9 (1875). 416-424.

[11] Wilkins, in his ed. of Hor. *Epist.*, takes the expression as a general one, but he does not consider *imus* a localizing adjective. To strengthen the theory of indefiniteness, it is worth noting that Horace has used not a present tense, but a future which seems almost equivalent to a potential subjunctive. Bennett, *Syntax of Early Latin*, 1.44 f. (Boston, 1910), does not recognize such a future; Blase, *Hist. Gramm. der lat. Spr.* 3.119 (Leipzig, 1903), terms it the "Futurum der Wahrscheinlichkeit."

[12] Cp. Hor. *Carm.* 1.34.12 f., *ima summis / mutare*; 3.1.15, *insignis et imos*.

illustration of the *faber* is meant to elucidate. It reads:

> In vitium ducit culpae fuga, si caret arte.

Considering the close connection of the lines, it can scarcely be doubted that *imus* merely repeats the thought of *caret arte*. If further evidence is demanded, it may be noted that Horace uses the adjective in a similar signification in the *Ars Poetica* 378, where he maintains that "if a poem swerves in the least degree from excellence, it sinks to insignificance" ("si paulum summo decessit, vergit ad imum").

It is quite possible, of course, to read into the phrase, *faber imus*, a reference to the sculptor's social position. Perhaps he *was* "humble" in rank, and so, through inefficient training, was unable to grasp a true artist's vision; but Horace had all due respect for the lowly born; at all events, in these disputed lines, his criticism certainly appears to be directed against lack of skill, not want of high birth. In our zeal for discovering hidden significations in words, we must not be carried away by an imagination like that of Acron, who reports that in the opinion of some people, *imus* means "short,"—he himself is inclined to think that "Imus was a certain statuary"!

When we turn to Martial, we find that his chief grievance against *fabri aerarii* was that the clatter of their mallets began at a very early hour in the morning, and disturbed the quiet all day long; we may infer, however, that they annoyed him no more than did chattering schoolmasters and sundry other noisy elements of the great city.[13]

Ferrarii Owing to the absence of *ferrarii* from Plutarch's list of the first industrial corporations at Rome,[14] we may suppose that the metal implements used by the primitive Romans were ordinarily cast from bronze.[15] But with the increasing use of iron for military, agricultural, and other common implements,[16] much work that had previously been done by *aerarii* passed into the hands of *ferrarii*. Because of the nature of their output, A. Typaldo-Bassia is inclined to grant a special dispensation for social recognition to certain workers in iron and bronze. Although he believes that

---

[13] Mart. 9.68; 12.57; cp. Juv. 7.222 f.

[14] There was a *conlegium fabrum ferrarium* at Rome at the beginning of the first century of our era; cp. CIL. 6. 1892; Waltzing 2.122.

[15] Cp. Liv. 1.43.2; Wezel 15; Marquardt 2.392 f.; Peet 492 f., 495-497, 510. For comparison with Greece, cp. Lang, *Early Uses of Bronze and Iron*, Class. Rev. 22 (1908). 47.

[16] Cp. Sen. *Herc. F.* 930 f.; Petron. 108; Mart. 14.36; Juv. 3.309-311; 15.165-168. Iron became so common that *ferrum* and *ferramenta* seem sometimes to be employed generically for "implements" or "utensils," even when the material used is bronze.

Rome looked with disfavor upon most of the industrial and commercial classes, and gave little consideration to the working man, because it was essentially a military nation; yet for this very reason, he maintains: "Toutefois il existait un genre d'ouvrage qui était au-dessus du préjugé traditionnel et national; je veux parler de la construction des machines de guerre et de la réparation des armes ou projectiles."[17] He offers only the *a priori* argument to support his claim. Perhaps he had in mind Mamurius Veturius,[18] who was said to have been one of the greatest artificers of the time of Numa; he was held in such high honor for producing shields in exact imitation of the *ancile* which dropped from the sky that his name was perpetuated in the song of the Salii.[19]

A. Typaldo-Bassia's simple reasoning seems somewhat fallacious. On such grounds it would be possible to make a rather sweeping assertion, highly agreeable to the present thesis, exempting from varying degrees of social disdain not only all *fabri* who followed the army, but *coriarii* who tanned leather for military purposes, *sutores* who made *caligae* for soldiers, *infectores* who dyed military *vexilla* and *tunicae*, *fullones* who pressed triumphal robes and military apparel, and all other workers directly or indirectly connected with military service. So far as the satirists are concerned, however, there is no evidence strong enough to gainsay the conclusion that *fabri aerarii* and *ferrarii* as a class

[17] Typaldo-Bassia 3.

[18] It is true that this name has been branded as eponymous to explain the words *Mamuri Veturi* in the hymn of the Salii, which, according to Varro *Ling.* 6.49 (45), really mean *veterem memoriam*; but see Wezel 16 f. for a review of the various theories and a defense of the *nomen*. The objection that Prop. 4(5).2.59-64 says that this Mamurius made the wooden statue of Vertumnus in the Vicus Tuscus, while Varro distinctly states that the Romans had no anthropomorphic representations of the gods for 170 years (cp. Plut. *Numa* 8; Aug. *Civ.* 4.31; Wissowa 32), is not unanswerable. Propertius seems to be the only authority for his assertion, and he was possibly not an art connoisseur. Since many *aerarii* in his day were statuaries, it would be quite natural for him to assign this ancient image to the earliest *faber aerarius* of his knowledge. The god was an Etruscan deity; it was in all probability the workmanship of an Etruscan, it was possibly even made in Etruria and brought to Rome by those who settled in the Vicus Tuscus. As it is now believed by some scholars that this settlement was composed of the workmen who had gone to Rome to build the temple of Jupiter Capitolinus in the time of Tarquinius Superbus (cp. Platner 172), in referring the statue to this period, there would be no conflict with Varro's statement. Propertius's words of commendation for the *formae caelator aenae* are suggestive of an appreciative attitude shown toward *aerarii* of his own day, or at least for those among them who were also skilled engravers and embossers, *caelatores* (see p. 8).

[19] Dionys. Hal. 2.71; Ov. *Fast.* 3.367-392; Plut. *Numa* 13.

were free from what Typaldo-Bassia chooses to call "traditional[20] and national prejudice." The irony in Juvenal's lines, for instance, which represent Vulcan as a blacksmith, attending a banquet of the gods before he thinks to wipe his arms, which are black from the smutty workshop,[21] is directed primarily against the extravagance, perfidy, and irreverent religious tendencies of the poet's day; nor is it fair to say that the details, which seem purely conventional, heighten the sarcasm, unless the same is said also in the case of Saturn and his *agrestem falcem*.[22] As a matter of fact, in his famous Satire on the Vanity of Human Wishes, Juvenal implies that the blacksmith's life, though confessedly humble, might offer in the end more happiness and contentment than an illustrious career. It was *dis adversis*, he says, that Demosthenes abandoned his father's smithy to attend a school of oratory.[23] It is beside the point to argue that Demosthenes' father was not a blacksmith, but a wealthy manufacturer of swords, who derived three-fourths of his fortune from other channels. At all events he was called "the cutler," and the term was applied in derision.[24] Whether Juvenal accepts the tradition as fact or metaphor is immaterial. The significant thing is that the poet criticizes and reproves the unsocial spirit that had been dominant at Athens in the fourth century B C., a spirit which had been strongly reflected in republican Rome, and was still struggling to exist; but he praises the more liberal attitude of mind that was beginning to assert itself under the Empire in the belief, on the part of many, that a life of worth and satisfaction could be found amid humble pursuits.[25]

II	III	IV
Argentarii	Aurifices	Caelatores

A distinction must be made between *argentarii*, who acted in the

[20] That it was not traditional, Wezel 32 maintains with ample proof, and he is convinced: "Opificium illa aetate (i.e., Numae) aequalibus non sordidum neque abiectum visum esse, cum rex studia opificum legibus ac beneficiis adiuvaret."

[21] Juv. 13.44 f.; cp. *schol.* and Munro in Mayor's ed. of Juv.; Lucian *Deor. Dial.* 5.4 f. For those who understand from Juvenal's verses that Vulcan is not a cupbearer, but a guest, there is even less occasion for ascribing extra acidity to the poet's qualifying phrase; for is it not probable that his picture of society among the gods is based upon actual social conditions existing in his own day?

[22] Juv. 13.39 f.

[23] *Id.* 10.129-132.

[24] For a collection and discussion of the literary evidence on this subject, see Arnold Schäfer, *Demos. u. seine Zeit*, 1.261-273 (Leipzig, 1885); Mayor on Juv. 10.130.

[25] See pp. 86 f., on *fabri* in general.

capacity of brokers,¹ and *fabri argentarii*, who worked as silversmiths.²
With the former, we are not concerned here;³ to the
Argentarii latter, the satirists appear to have made no direct reference. They are not even mentioned in the list of guilds which Plutarch ascribes to Numa;⁴ silver, it seems, was not in common use at Rome in the early days of the city,⁵ and it is possible that at first, as was sometimes the case later, the crafts of *argentarii* and *aurarii* were combined.⁶ Waltzing finds no datable evidence for a special corporation of silversmiths until the beginning of the third century of our era.⁷ Among sepulchral inscriptions there is one commemorating a *faber argentarius* who had served as a *magister vici*.⁸

Goldsmiths, *aurifices, fabri aurarii*, could boast greater antiquity for their trade; for they were represented in the so-called guilds of Numa.⁴
Aurifices A *conlegium aurificum* still existed in Rome in the time of Augustus;⁹ and there is evidence for separate corpora-
Aurarii tions of *anularii*¹⁰ and *brattiarii* (=*bractearii*) *inaurati*¹¹ in the city. According to an inscription of Amsoldingen near Thun, two *aurifices*, father and son, had been members of a body of carpenters and builders, *corporis fabr. tignuariorum*; the father had held all the offices of the association.¹² At Pompeii, *aurifices universi* figure in the wall *graffiti*, designating their preference for candidates in the municipal elections.¹³ Yet goldsmiths receive no special notice from any of the satirists of the Empire except Martial, and he merely accuses them of contributing to the city's incessant din from a very early hour in the day.¹⁴

¹ Sen. *Contr.* 1 *praef.* 19; Suet. *Nero* 5; Acron on Hor. *Sat.* 1.6.85; CIL. 6.9178, 9181, 9186.
² CIL. 6.2226, 9209, 9390-9393.
³ Cp. Waltzing 2.114 f.; Blümner, Müller's *Handbuch* 4.2.2.651-656.
⁴ See p. 1.
⁵ Cp. Darem.-Saglio 1.1.410; Blümner, *op. cit.*, 265, 392.
⁶ Cp. CIL. 6.9209; 11.3821.
⁷ Waltzing 2.111.
⁸ CIL. 6.2226.
⁹CIL. 6.9202; Waltzing 2.111; 4.8 f. (Waltzing makes a careless mistake here, quoting Pliny *Nat.* 34.1, which refers to *aerarii*; see p. 1, n. 2.).
¹⁰ CIL. 6.9144.
¹¹ CIL. 6.95; cp. 6.9210, 9211.
¹² CIL. 13.5154.
¹³ CIL. 4.710.
¹⁴ Mart. 12.57.9 f.

Expert workers in gold and silver, or even bronze,[15] who were skilled in basso-relievo chasing, were called *caelatores*. The pages of the satirists contain a few chance allusions to them. In the time of Martial and Juvenal, chased gold and silver plate was in such demand

Caelatores that even the man in slender circumstances desired to find among his skilled slaves a *curvus caelator* attentively stooping over his work.[16] Choice old pieces made by famous Greek *caelatores* of the fifth and fourth centuries B. C., such as Myron,[17] Mentor,[18] Mys,[19] and Polycleitus,[20] or even Phidias,[21] Scopas,[22] and Praxiteles,[22] were especially sought; but in lieu of these, specimens of modern workmanship would suffice,[23] especially when they were so delicately wrought as to admit the possibility of being passed off as genuine antiques.[24] The shops of goldsmiths and silversmiths were to be found especially along the Sacra Via[25] and in the Saepta.[25]

Kühn believes that the occupation of those who worked with the precious metals was not popular among *ingenui*; his lists of inscriptions concerning them show freedmen in the majority.[26] It may be argued that freeborn Romans were disinclined to engage in the handicraft of the goldsmith and silversmith because of no feeling of contempt for it, but because of inferior ability; not possessing the innate artistic instincts of the sons of Hellas, they would easily be eclipsed by the Greek *liberti* among their competitors. As a class, these craftsmen must have been men of no mean endowment and refinement; for their tasks required not only skill, but intelligence and artistic taste. They doubtless met with due respect at the hands of liberal minded Romans. Among inscriptions we find the following appreciative epitaphs dedicated to two of their number:

[15] Prop. 4(5).2.61; Ov. *Fast.* 3.381. See p. 5, n. 18.
[16] Juv. 9.145.
[17] Mart. 4. 39. 2; 6. 92.2; 8.50(51).1.
[18] *Id.* 3.40(41).1; 4.39.5; 8.50(51).2; 9.59.16; 11.11.5; 14.93.2; Juv. 8.104.
[19] Mart. 8.34.1; 8.50(51).1; 14.95.
[20] *Id.* 8.50(51).2.
[21] *Id.* 3.35.1; 4.39.4; 10.87.16.
[22] *Id.* 4.39.3.
[23] Cp. *Clodiana, Furniana, Gratiana,* Plin. 33.139; Mart. 4.39.6; CIL. 6.9222.
[24] Mart. 8.34.
[25] CIL. 6.9207, 9212, 9214, 9221, 9393. See pp. 69 f. CIL. 6.9208 names an *aurifex extra Porta Flumentan.*
[26] Kühn 45-47.

ROMAN CRAFTSMEN AND TRADESMEN OF THE EARLY EMPIRE 9

L. VETTIVS NYMPHIVS AVRIFEX . V.A . XVII ET . TE . TERRA . PRECOR . LEVITER SVPER . OSSA . RESIDAS SENTIAT . VT . PIETAS . PRAEMIA QVAE . MERVIT ET . QVICVMQVE . SVIS . SINCERE PRAESTAT . HONOREM FELICEM . CVRSVM . PERFERAT AD . SVPEROS[27]	D . M M . CANVLEI ZOSIMI VIX . ANN . XXVIII FECIT PATRONVS . LIB BENE . MERENTI HIC . IN . VITA . SVA . NVLLI . MA LEDIXIT . SINE . VOLVNTATE PATRONI . NIHIL . FECIT MVLTVM . PONDERIS AVRI . ARG . PENES . EVM SEMPER . FVIT . CONCVPIIT . EX . EO NIHIL . VMQVAM . HIC . ARTEM . CAELA TVRA . CLODIANA . EVICIT . OMNES[28]

V

CAUPONES

After his journey to Brundisium with the envoys whom Octavian was sending to Antony, Horace may have cherished some pleasant memories of days spent in comradeship with such literary lights as Plotius, Varius, and Virgil, and with the eminent embassadors Cocceius, Fonteius Capito, and Maecenas,[1] but there were no cheering recollections of luxurious hotel accommodations. In his day, as during the Republic and on into the Empire, *cauponae*[2] and *tabernae deversoriae*,[3] whether wayside inns or the best that lavish Capua offered, were apparently very undesirable places, to be endured for temporary shelter, but to be avoided whenever possible.[4] On the much travelled Appian Way the poet found at Aricia only a *hospitium modicum*;[5] and at Forum Appii,

[27] CIL. 6.9204.
[28] CIL. 6.9222.
[1] Hor. *Sat.* 1.5.
[2] *Ib.* v.51; *Epist.* 1.11.12; 1.17.8; Plin. *Nat.* 9.154; Gell. 7.11.4, *cauponula*.
[3] Varro *Rust.* 1.2.23; Cic. *Orat.* 2.234; *Epist.* 6. 19.1; 7.23.3; *Cato* 84; Liv. 45.22; Petron. 9, 15, 19, 81 f., 124, cp. 80, *humilis taberna*.
[4] Cic. *Epist.* 6.19.1; *Cato* 84; Hor. *Epist.* 1.11.11 f.; 1.17.7 f.
[5] Hor. *Sat.* 1.5.1 f. Cp. Cic. *Orat.* 2.234; *Cato* 84; Petron. 85 f., 91. Other terms for "lodgings" are *cenaculum* (Petron. 38; Hor. *Epist.* 1.191, the word apparently means "boarding house"), *insula* (Petron. 98), *stabulum* (*Id.* 6, 8, 16, 79, 97). *Insula* may be applied also to a block of houses or "tenement" (Mart.4.37.4; Suet. *Iul.* 41; *Tib.* 48; *Nero* 38; Tac. *Ann.* 6.45; 15.43).

caupones maligni;[6] while in the important city of Beneventum, the hostelry was so old that it furnished ready fuel for the rapid spread of a fire, which was caused by a mishap when the bustling host was preparing a hurried meal; and it is to be noted that the first efforts of hungry guests and terrified slaves were directed toward saving the dinner, meager though it was![7]

Wealthy and influential travellers were forced to put up with such unsatisfactory quarters, only when they had no country estates of their own, and when the establishments of friends or houses of call were either inaccessible or inconvenient. Villas were doubtless quite numerous along the various highways from Rome; and according to Cicero: "Semper . . . boni assiduique domini referta cella vinaria, olearia, etiam penaria est, villaque tota locuples est, abundat porco, haedo, agno, gallina, lacte, caseo, melle."[8] On the embassy to Brundisium, Maecenas's party was entertained at Formiae by Fonteius Capito,[9] and by Murena, who was probably the prime minister's brother-in-law;[10] and the inns of the Caudine Forks were scorned for the *plenissima villa* of Cocceius.[11]

A less commodious and bountiful home, termed by Horace a *villula*, had opened its doors to them below Sinuessa.[12] This was presumably one of those rural houses which, though possibly belonging to private individuals, were appointed by the government to supply the needs of magistrates and public officials who were travelling on state business. Their hosts were called *copiarii*, or *parochi*, a word which, as Porphyrio notes, was derived from the Greek παρέχειν.[12] To prevent extortion on the part of guests, the *Lex Iulia de Repetundis*[13] had prescribed that road houses were bound to furnish only necessary shelter and food for

---

[6] Hor. *Sat.* 1.5.3 f.

[7] *Ib.* vv. 71-76.

[8] Cic. *Cato* 56. Cp. Mart. 3.58 for a description of Faustinus's splendid rural estate at Baiae; the epigram shows the contrast between an artificial villa and a well stocked country place.

[9] Hor. *Sat.* 1.5.37 f. So far as the evidence in the lines is concerned, there seems to be no occasion to infer with Verrall and many editors that Murena himself was absent. The obvious interpretation seems to be that Capito entertained the envoys at dinner, but that they spent the night at Murena's villa.

[10] Cp. Dio 54.3. On the identity of Murena, cp. Verrall, *Studies in the Odes of Horace*, 16-18, 83-86 (London, 1884).

[11] Hor. *Sat.* 1. 5. 50-70. On Roman Hospitality, cp. Miller 29-33.

[12] Hor. *Sat.* 1.5.45 f. and Porph.

[13] Cp. Tyrrell and Purser, *Correspondence of Cicero*, 3.295 f. (London, 1890).

man and beast. From all accounts, the bare letter of the law was observed.[14]

Landowners who had country estates bordering upon roads frequented by travellers, were urged by Varro to build *tabernae deversoriae* as adjuncts to their villas, in order to dispose of their produce.[15] If the advice was taken, the inns were probably put in charge of slaves or freedmen.[16] Martial's clause, "Non segnis albo pallet otio caupo," which is descriptive of Faustinus's villa at Baiae, may refer to the practice.[17]

Cauponae at Rome were included by Martial among the shops that had become a nuisance to pedestrians; flagons chained to their door posts blocked the sidewalks.[18] The city inns appear to have been primarily wine shops;[19] they must, of course, have furnished meals to guests; but they are to be clearly discriminated from cheap restaurants, *popinae*, where hot food and drinks were served to chance customers.[20] *Tabernae vinariae* were evidently considered more necessary and less disreputable, and they were probably visited more frequently and openly by men who laid claims to respectability. Martial confesses that they were essential to his own happiness.[21] Their greatest patronage, however, was doubtless from slaves, but the domestic servants, *libarii*, *archimagiri*, *carptores*, whom Juvenal enumerates among the scandalmongers that gathered with *caupones* before dawn, seem to have been considerably less objectionable characters than the murderers and other reprobates whom he assigns to *pervigiles popinae*.[22]

The reputation of *caupones* themselves is readily seen from the open characterizations and covert insinuations of the satirists and others.

[14] Cic. *Att.* 5.16.3; Hor. *Sat.* 1.5.46, cp. 2.8.36 where *parochus* is used jocosely for a miserly host; Marquardt 1.199.

[15] Varro *Rust.* 1.2.23.

[16] Cp. Petron. 61.

[17] Mart. 3.58.24 *Caupo*, however, is suspicious here; out of nine cases this is the only place in which Lindsay adopts the spelling *caupo* instead of *copo*; of the MSS. B^A has *carbo*, C^A *capo*.

[18] Mart. 7.61.5, 9. See p. 61.

[19] Mart. 1.26.9; 2.51.3.

[20] Hor. *Epist.* 1.14.21, 24; Mart. 7. 61.5, 8 f. Some *cauponae* may have degenerated into *popinae*; perhaps in small towns, like Pompeii, the latter were near inns or connected with them, cp. Mau-Kelsey, 400-404. See pp. 15 ff.

[21] Mart. 2.48.1; cp. Suet. *Claud.* 40.1.

[22] Juv. 8.158, 172-176; 9.107-112; 11.80 f.; cp. Hor. *Epist.* 1.14.24.

Horace's *sedulus hospes*,[23] *perfidus caupo*,[24] and *cauponibus malignis*[25] testify to the innkeeper's dishonesty and maleficence, and to his obsequious or enticing manner.[26] Virgil's *Copa* gives an attracting invitation to a pleasant arbor and a babbling brook, to roses, violets, lilies, chestnuts, and luscious fruits; but these very charms lure to temptation, and the Syrian hostess herself is guileful. In Petronius, inns and lodging houses are places of riot and houses of ill repute;[27] and Juvenal exposes the domestic troubles of a city tavern-keeper whose wife was the type of woman that resorted to the fortune tellers of the Circus and the Agger.[28] Martial deemed *caupones* lazy[29] and deceitful. Among his indictments, we find:

> Callidus inposuit nuper mihi copo Ravennae;
> cum peterem mixtum, vendidit ille merum.[30]

The poet has adroitly prepared his readers for the justice of *callidus* and *inposuit* in the neighboring couplet:

> Sit cisterna mihi quam vinea malo Ravennae,
> cum possim multo vendere pluris aquam.[31]

But pure wine was very rare, it seems, in a country where water was plentiful; in fact, hints the epigrammatist, honesty in tavern-keepers was prevented by fate; for

> Continuis vexata madet vindemia nimbis:
> non potes, ut cupias, vendere, copo, merum.[32]

Petronius also, in jocund vein, notes the hand of destiny and makes Trimalchio in one of his garrulous outbursts offer the suggestive information that *caupones* were born under Aquarius, along with "cabbage-

---

[23] Hor. *Sat.* 1.5.71.

[24] *Ib.* 1.1.29. This reading seems absolutely irrelevant to the context. The poet has twice grouped *mercator* and *miles*, *agricola* and *iuris consultus*, and now returning to them for the third time, strangely substitutes *caupo* for lawyer. *Causidicus*, or a circumlocution of kindred meaning, would still be consistent and would satisfy the objection that jurisconsults received no money. I am, therefore, inclined to adopt one of the various emendations, preferably Müller's *perditus hic causis*; cp. Orelli-Mewes' ed. of Hor.

[25] Hor. *Sat.* 1.5.4.

[26] Cp. Cic. *Cluent.* 163; Virg. *Copa*; Plut. *De Vit. Pud.* 8; Juv. 8.159-162.

[27] Petron. 9, 15 ff., 79 ff., 91 ff.; cp. Virg. *Copa* 1-6, 31-34.

[28] Juv. 6.588-591; cp. Petron. 61. By the Agger is meant, presumably, the Rampart of Servius Tullius between the Porta Collina and the Porta Esquilina, cp. Lanciani, *Ruins and Excavations*, 62 (New York, 1897); Platner 114. See p. 76.

[29] Mart. 3.58.24, but see n. 17.

[30] *Id.* 3.57.

[31] *Id.* 3.56.

[32] *Id* 1.56.

heads," *cucurbitae*.[33] Martial adds further accusation in a distich which was intended to be attached to barley when presented as a gift:

> Mulio quod non det tacituris, accipe, mulis.
> haec ego coponi, non tibi, dona dedi.[34]

Friedländer interprets this to mean that the muledriver sells to the innkeeper at a very low figure the barley apportioned to him for his mules. But the epigram is apparently not addressed to the muleteer, and the clause, "mulio quod non det," merely describes the poor quality of the *hordeum*. Pliny says that bread made from barley was in favor among the ancients but that in his time, "quadrupedum fere cibus est."[35] Perhaps, then, the inference in the lines quoted above is as follows: "The barley which I am giving you is of such poor quality that the muleteer would not give it to his dumb animals, but the innkeeper will have no scruples against using it, and so you will turn my gift over to him."

After such testimony, we cannot but assign *caupones* to a very low rank in the social scale. Martial was amazed to find that a cobbler and a fuller had achieved the degree of success that permitted them to exhibit gladiatorial shows, but he intimated that the limit would be reached when these should be given by a *caupo*.[36]

VI Centonarii

This subject is treated with *Tignarii*, in XXV, page 80.

VII Cerdones

This subject is treated with *Sutores*, in XXII, page 54.

VIII
Coci*

At the time when the Romans were first divided into guilds according to their various occupations,[1] no *collegium cocorum*, it appears, was

[33] Petron. 39.
[34] Mart. 13.11.
[35] Plin. *Nat.* 18.74.
[36] Mart. 3.59.

*For a discussion of the two spellings *coqui* and *coci*, cp. Harcum 7. The present section was written before the appearance of Dr. Harcum's dissertation on *Roman Cooks*. References to her monograph have been inserted at various points; but since she has concerned herself chiefly with domestic cooks, and has either overlooked or disregarded cooks in trade, there is no serious overlapping.

[1] See p. 1.

instituted; for, as Waltzing suggests, cookery was merely a household occupation.[2] Like baking, it would naturally be the work of women[3] or of domestic servants; indeed there do not seem to have been even special slaves for this purpose at first. Apparently general domestics prepared the ordinary meals; but by the time of Plautus, at least, there is evidence indicating that professional cooks could be hired from the Forum;[4] gradually, trained *coci* were to be found in private households. We have it on Livy's authority that slave cooks had been considered very worthless chattels by the ancient Romans both in value and in usefulness; but, he adds, with the introduction of extravagant and luxurious tastes upon the return to Rome of the victorious army from Asia, even meals were prepared with greater care and expense; then cooks became valuable, and what had hitherto been a menial service began to be considered an art.[5]

The *cocus* and his *ars* continued to rise in money value as the desire for delicacies increased. The elder Pliny offers a delightful "fish story" apropos of this. After noting that complaint had formerly been raised because a cook cost as much as a horse, he solemnly avows, "but now cooks are purchased at the price of three of these, and it takes the cost of a cook to buy a fish!"[6] Juvenal counts the *cocus*, and the *structor* associated with him,[7] among the costly luxuries of an establishment that boasted every extravagant and elaborate appointment.[8] According to Martial, it even came into vogue for those who were especially *gulosi* to secure as cooks slaves of marked beauty of hair and feature, such

[2] Waltzing 1.67. Later corporations of cooks were apparently *collegia domestica*, cp. Waltzing 1.215, 346; 2.148; 4.154 f.; Harcum 79 ff. CIL. 11.3078 records a possible exception, but this is referred to cooks only through a conjecture, cp. CIL. *l.c.*, note.

[3] Cp. Plin. *Nat.* 18.107.

[4] Plaut. *Aul.* 280 f.; *Merc.* 697; *Pseud.* 790; Plin. *Nat.* 18.108; Harcum 15-18, 58-60.

[5] Liv. 39.6. The *Cena Trimalchionis* of Petron. abounds with examples of the cook's skill (*e.g.*, 47, 49, 74); Daedalus, whose very name proclaimed his ingenuity, could serve an innocent pig as fish, woodpigeon, bacon, turtle dove, and fowl (69 f.; cp. Mart. 11.31.11-14). A cook-book entitled *Apici Caeli de re coquinaria* is still extant. This collection of recipes of every variety was compiled possibly about the third century after Christ; on the question of its authorship, cp. Teuffel-Schwabe-Warr2.§ 283.2; Schanz 506 f. Cp. Harcum 9-14, 47 f.

[6] Plin. *Nat.* 9.67. Cp. Harcum 51-57.

[7] On the cook's assistants, cp. Harcum 69-77.

[8] Juv. 5.120-123; 7.184 f. and *schol.*; 11.136-141; cp. Mart. 10.48.15.

as were far better fitted to be cupbearers. The poet makes protest in the following indignant epigram:

> Quis, rogo, tam durus, quis tam fuit ille superbus
> qui iussit fieri te, Theopompe, cocum?
> hanc aliquis faciem nigra violare culina
> sustinet, has uncto polluit igne comas?
> quis potius cyathos aut quis crystalla tenebit?
> qua sapient melius mixta Falerna manu?
> Si tam sidereos manet exitus iste ministros,
> Iuppiter utatur iam Ganymede coco.[9]

Cooks were not engaged in domestic service only. Especially during the Empire, many *coci* conducted eating houses, *popinae*, and swelled the ranks of culpable shopkeepers who blockaded the public sidewalk in front of their *tabernae*, until they were restrained behind their thresholds by a law of Domitian.[10]

Restaurants are represented as hot, grimy, dirty, greasy places.[11] Before some of them probably hung canvas awnings or curtains on which the bill-of-fare or business sign was inscribed; this we may infer from the following lines of Juvenal:

> Lateranus ad illos
> thermarum calices inscriptaque lintea vadit[12]

This passage may also be cited as evidence that cafés were sometimes located near baths, or were even connected with them.[13] Corroborating testimony is furnished by Martial, who tells us that Syriscus squandered ten million sesterces, "in sellariolis vagus popinis / circa balnea quattuor."[14] The force of *sellariolis* is plainly seen from the closing lines of the epigram:

[9] Mart. 10.66; cp. 12.64.
[10] *Id*. 7.61.8 f.; cp. 1.41.10. See p. 61.
[11] Hor. *Sat*. 2.4.62; *Epist*. 1.14.21; Mart.5.44.7-10; 7.61.8; Juv. 11.81.
[12] Juv. 8.167 f. and *schol*.; cp. Mart. 1.117.11; Friedländer-Magnus 1. 291 f. This is the interpretation adopted by Mayor and Friedländer in their editions of Juv. Lewis, commenting upon the same passage in his own edition, adduces strong evidence in support of his argument that *lintea* refers to curtains of brothels, but it seems to me that the lines in question are so closely connected with what precedes and follows that verse 168 is but a circumlocution for *popinas* and *popina* of verses 158 and 172. Indeed *Syriae* in 169 may be a not unintentional repetition of *Syrophoenix* of 159 f.: "Lateranus actually frequents cook-shops and yields to the allurements of Oriental hosts, although he is of ripe age to go to the Orient to battle for his country."
[13] Wright, in his ed. of Juv., sees here an opportunity to compare *popinae* with the *thermipolia* of Plaut. *Curc*. 292; *Rud*. 529; *Trin*. 1013.
[14] Mart. 5.70; cp. Sen. *Epist*. 122.6, Quint. 1.6.44, Mart. 17.70, although these may refer to *cauponae*.

> O quanta est gula, centiens comesse!
> quanto maior adhuc, nec accubare!

The gourmand's gluttony is shown to have been all the greater because he indulged it in ordinary cook-shops, where people sat upon common stools,[15] instead of amid the refinements of a respectable dining-room, where he would recline upon a couch.

Popinae offered for sale wine,[16] cooked meats, and vegetables, including the richest and rarest delicacies.[17] They seem also to have sent out hawkers, called *coci*[18] or *popinarum institores*[19] who peddled smoked sausage about the streets. Various emperors attempted to check intemperance by sumptuary laws. Tiberius forbade the sale of any cooked food, even pastries.[20] Claudius, however, who had tenants of his own engaged in the business, disregarded this ban and removed the aediles' supervision.[21] Nero passed a new law permitting the sale of no cooked food except vegetables, *legumina aut holera*.[22] His regulation was evidently not enforced,[23] for Vespasian had to make a similar one.[24] He appears to have met with as little success, since we have only to turn to Juvenal to read of the rich viands that were obtainable in restaurants under Hadrian.[25]

Men who frequented cook-shops were usually of the most degraded type. Horace's phrase, *obscuras humili sermone tabernas*,[26] may well be applicable here. In Petronius, we find Eumolpus on the point of going the rounds of the eating-houses to search for a lost slave.[27] According to Martial, gamblers might be sheltered with their illicit pleasure in a secluded *popina* until they were spied and dragged away by the aedile.[28] Among other regular patrons were numbered criminals and the lowest

---

[15] See the illustration in Darem.-Saglio 1.2.973, fig. 1257, reproducing the wall painting in a Pompeian café.

[16] Mart. 5.84.5; *Juv.* 8.162, 168, 177.

[17] Hor. *Sat.* 2.4.58-62; Mart. 5.44.7-11; Juv. 11.81; Suet. *Nero* 16.1.

[18] Mart. 1.41.9 f. See p. 23.

[19] Sen. *Epist.* 56.2.

[20] Suet. *Tib.* 34.1.

[21] *Id. Claud.* 38.2; but Dio 60.6 says that Claudius, too, forbade the sale of dressed meats and hot water.

[22] Suet. *Nero* 16.2.

[23] Cp. *Id. Vitel.* 13.3.

[24] Dio 66.10.

[25] Juv. 11.81.

[26] Hor. *Epist.* 2.3.229. See p. 72.

[27] Petron. 98.

[28] Mart. 5.84.3-5.

class of slaves: *percussores, nautae, fures, fugitivi, carnifices, fabri sandapilarum, galli*,[29] *squalidi fossores*,[30] *mediastini*.[31] The application of the term *popino* to any one carried gross insult. Lenaeus the freedman of Pompey hurled the word at Sallust, when he was enraged at the historian for some slighting remarks about his master.[32] Yet rumor connected with cabarets even noble names. As we have seen, one of the aristocratic Laterans, possibly Plautius Lateranus who was consul elect in 65 A. D. but was put to death by Nero before he could enter upon his office,[33] was accused by Juvenal of repeated visits to a *pervigilis popina*, where he was accorded a most cordial welcome by an enticing host and barmaid.[34] Claudius apparently had an early predilection for similar resorts.[35] Nero, it is said, disguised himself by a peruke or by a freedman's cap and entered cook-shops after dark.[36] Vitellius, too, when travelling, availed himself of cafés to allay his insatiable appetite.[37] Such habits in men of mature years left a deep and lasting blot upon their reputation, but indiscreet young nobles, it seems, might yield to temptation, and still hope to find many a lenient censor who would recall: "Fecimus et nos / haec iuvenes."[38]

When we note the low nature of restaurants and find the sentiment expressed that the kitchen was no place for even a slave if he be comely, we cannot but conclude that *coci* met with little esteem. They were, perhaps, almost at the bottom of the social order. The evidence at hand indicates that the majority of them were slaves, or at best, freedmen.[39] It is probable that they were seldom recruited from the ranks

[29] Juv. 8.171-178.
[30] *Id*. 11.80.
[31] Hor. *Epist*. 1.14.14, 21.
[32] Suet. *Gramm*. 15; cp. Hor. *Sat*. 2.7.39.
[33] Tac. *Ann*. 15.49.3; 15.60.1 f.; cp. 11.30.3; 11.36.5; 13.11.2.
[34] Juv. 8.158-162.
[35] Cp. Suet. *Claud*. 38.2; 40.1 (but see p. 11).
[36] *Id. Nero* 26.1.
[37] *Id. Vitel*. 13.3.
[38] Juv. 8.163 f.; cp. Suet. *Claud*. 16.1.
[39] Petron. 36, 38, 47 ff., 68, 70, 74 f.; Mart. 1.50; 3.13; 3.94; 5.50.8; 6.39.6 f.; 6.61(60).8; 7.27.7 f.; 8.23; 9.81.3 f.; 10.48.15; 10.66; 11.31.11-14; 13.10, 52, 70; 14.220; Juv. 5.120-123; 7.184 f.; 9.107-110; 11.136-141. These include allusions to *archimagiri, carptores, structores*. Most of them appear to refer to domestic *coci* or their assistants; it is possible that many household cooks after being emancipated, established public eating-houses. Cp. Harcum 20, 62-64, 67 f., and chapters 4 and 5 on the nationality and the names of cooks.

of *ingenui*,⁴⁰ nor did they come, as a rule, from a high class of slaves. One "excellent cook,"⁴¹ a freedman, is on record as having been prominently connected with the cult of the Emperor at Alba Fucens. He apparently served both as *dendrophorus*⁴² and as *sevir Augusti*;⁴³ such distinction for a *cocus* was doubtless very rare. Petronius brands all occupants of the kitchen as ill-smelling;⁴⁴ a peddler cook is described as a common *verna*, base in trade and character;⁴⁵ and the attributes applied to cafés (*immunda, uncta,*⁴⁶ *sellariola, arcana, nigra,*⁴⁷ *pervigilis*,⁴⁸ speak little for the refinement of their hosts.

IX
Coriarii

If we refer again to the *collegia* ascribed to Numa by Plutarch, we find recorded among them that of tanners, *coriarii*.¹ To have merited this special recognition they must early have conducted an extensive and important business; it is fair to assume that they became increasingly necessary and useful in the state.

Our investigations have revealed only one significant passage which bears directly upon tanners. This is in Juvenal's fourteenth Satire in which the poet deplores the avaricious spirit that leads a father to inculcate a desire for gain in his son, both by example and by precept. To this end, says the satirist, he urges the boy to join the ranks of lawyers, warriors, or merchants, and finally exclaims:

> Nec te fastidia mercis
> ullius subeant ablegandae Tiberim ultra,
> neu credas ponendum aliquid discriminis inter
> unguenta et corium.²

⁴⁰ But cp. CIL. 11.3078, and Waltzing 1.89; 4.87. See n. 2.
⁴¹ CIL. 9.3938.
⁴² See p. 82.
⁴³ See p. 96.
⁴⁴ Petron. 2, 70. On the characteristics and social position of domestic cooks, cp. Harcum 49 f., 62-68.
⁴⁵ Mart. 1.41.9 f.
⁴⁶ Hor. *Sat.* 2.4.62; *Epist.* 1.14.21.
⁴⁷ Mart. 5.70.3; 5.84.4; 7.61.8.
⁴⁸ Juv. 8.158.
¹ This is the interpretation of Plutarch's σκυτοδεψῶν (*Numa* 17) accepted by Wezel 25, Marquardt 2.392, Waltzing 1.63, and Blümner in Müller's *Handbuch* 4.2. 2.591 f. Mommsen, however, in his *Hist. of Rome*, 1.249, apparently translates it *fullones* in spite of its derivation; so Blümner in his earlier work, *Thätigkeit*, 110. See p. 1.
² Juv. 14.201-204.

These lines, though few in number, are replete with meaning. They point out that certain occupations, such as tanning, were relegated beyond the Tiber;[3] these had once been considered trades of last resort; but there was money in them, and one of the chief complaints against them was their offense to the olfactories. But the verses also admit a more subtle interpretation. As evidence for changing social conditions, they bear witness that commercialism in its steady progress was meeting less and less opposition, and that it had champions who considered aversion even to the trades of the Trastevere mere *fastidia*, not well grounded *odia*. Whether the reason was indomitable avarice, as Juvenal here maintains, or whether there was rather a complication of causes from which other social changes evolved under the emperors, the fact remains that commercialism was gaining enough recognition to be a cause for satire: fastidiousness in the choice of business occupations was disappearing, and there was beginning to be no distinction *inter unguenta et corium*.

X Dendrophori

This subject is treated with *Tignarii*, in XVII, page 82.

XI Fabri

This subject is treated with *Tignarii*, in XVII, page 78.

XII Ferrarii

This subject is treated with *Aerarii* in I, page 4.

XIII

Figuli

Pliny the Elder is authority for the statement that *figuli* composed the seventh corporation established by Numa.[1] He also shows that they continued to be indispensable under the Empire in spite of the lavish use of precious metals and crystal which great wealth permitted; for their energies were directed not only to the manufacture of crockery and vases of all kinds, but also to the making of bricks, tiles, and statuettes.[2]

---

[3] This was the case with the sulphur trade and other nuisances, such as peddlers, etc., cp. Mart. 1.41.3-5; 6.93.4. See pp. 22, 25.

[1] Plin. *Nat.* 35.159: "Septimum conlegium figulorum instituit." See p. 1, nn. 1 f.

[2] Hor. *Epist.* 2.3.21 f.; Plin. *l.c.*; Mart. 14.102,171, 178, 182; Juv. 4.133; 10.171; Marquardt 2.635-669; Blümner, *Tech.*, 2.5-7; Walters 2.279-555.

Modelers of clay, it appears, were not usually domestic slaves,[3] but worked independently or were employed in public or private *figlinae*.[4] It is interesting to note that Abbott's investigations on "Roman Women in the Trades and Professions" have revealed from a study of brick stamps that the brick business was largely in the control of women of prominent families.[5] Pliny names several towns in Italy, Asia Minor, Africa, and Spain which were famous for their potteries and brick plants.[6] Since the population of these places doubtless consisted primarily of *figuli*, we may conceive that there, at least, workers at the wheel and kiln were actively prominent in much of the life of the community. The satirists' references to craftsmen of this class are not oppressive with scorn or mockery, although Juvenal consigns them to the ranks of the humble.[7] The nickname Prometheus was apparently applied to them in jest rather than in ridicule; compare Martial's merry couplet for a clay figurine of a hunchback:

> Ebrius haec fecit terris, puto, monstra Prometheus:
> Saturnalicio lusit et ipse luto.[8]

Both Horace and Persius employed the metaphor of the potter's wheel to stand for that which was dearest to each, poetry and philosophy;[9] it seems unlikely that they would have drawn their figure from a source that was wholly despised or disdained.

XIV
FULLONES

When the first industrial colleges were instituted at Rome, it is possible that *tinctores*, dyers, were acting also in the capacity of *fullones*, fullers, since Plutarch does not mention the latter as forming a guild of their own.[1] Probably, however, as Wezel and Blümner[2] suggest,

---

[3] Cp. Juv. 4.134 f., "Sed ex hoc / tempore iam, Caesar, figuli tua castra sequantur," and Friedländer's comment.

[4] Marquardt 2.665-669.

[5] Abbott, *Society and Politics*, 98 f.

[6] Plin. 35. 160-173; cp. Walters 2.474-554.

[7] Juv. 10.171 f. shows the lowly degree of brickmakers by contrasting them with Alexander the Great: "Cum tamen a figulis munitam intraverit urbem, / sarcophago contentus erit."

[8] Mart. 14.182, cp. 176. Cp. also Juv. 4.133; Lucian *Prom. in Verb.* 2.

[9] Hor. *Epist.* 2.3.21 f.; Pers. 3.23 f.

[1] Plut. (*Numa* 17) uses the word βαφέων which, it is commonly agreed, means dyers. Mommsen, contrary to most scholars, interprets σκυτοδεψῶν as fullers. See pp 1; 18, n. 1.

[2] Wezel 25-27 thinks that he also has evidence for *carpentarii, lanarii coactiliarii*, and *lanii*; Blümner, Müller's *Handbuch* 4.2.2.593, names besides *fullones: lanii, piscatores, restiones*.

they were included in the ninth general division that represented various kinds of trades.³ At all events they must have been in business from very early days. Livy, for instance, speaks of a law of the fifth century B. C. which forbade political candidates to have their robes whitened with fullers' chalk when canvassing for office.⁴ Pliny cites the *lex Metilia . . . fullonibus dicta*, prescribing the technical processes to be employed.⁵ Waltzing agrees with Dirksen that this was a statute directed against the frauds of individual artisans.⁶ Pliny's addendum, "Adeo omnia maioribus curae fuere," implies that such measures, though probably necessary, were not frequently enacted in his own day, but that the industrial worker usually went his way unmolested.

According to Martial 14.51.2, "Non tam saepe teret lintea fullo tibi," fullers evidently cleaned linen in addition to woolen garments;⁷ in the first century, too, as in the twentieth, laundries were "hard on the clothes"; this is one of the instructive items to be gathered from the conversation at Trimalchio's dinner; for Seleucus, taking his fellow guests into his confidence, proclaims, "I do not bathe every day; for the bathman is a regular fuller, and water has teeth."⁸ Since fulleries were both tailoring and laundering establishments, in view of the foppery and extravagance of the first centuries of the Empire, when a man whose toga did not hang well was a laughing-stock,⁹ and when one prided one's self on the number of one's cloaks and syntheses and on the neatness of the plaits in one's attire,¹⁰ it would seem that fullers were on the high-road to wealth and success. Some evidently attained their goal; for Martial admits, though with no satisfaction, that a *fullo* had exhibited a gladiatorial show at Mutina.¹¹ The epigrammatist apparently felt no brotherly affection for *fullones*; their associations were distinctly

---

³ For the view that this ninth class contained, not *opifices*, but the "farmers and the rest of the citizens," cp. Mommsen, *De colleg.*, 29.

⁴ Liv. 4.25.13.

⁵ Plin. *Nat.* 35.197 f. The law probably belongs to the year 220 B. C., when L. Aemilius and C. Flaminius were censors, cp. Smith, *Dict. of Biog.*, 3.1359.

⁶ Waltzing 1.183.

⁷ On the fulling business, cp. Marquardt 2.527-530; Blümner, Müller's *Handbuch* 4.2.2.256; *Tech.*, 1.170-190.

⁸ Petron. 42.

⁹ Hor. *Epist.* 1.1.95-97.

¹⁰ Mart. 2.46; 2.57; 3.56.10; 4.66.3 f.; 5.79.

¹¹ *Id.* 3.59.2.

unpleasant to him; their vocation had an odorous connection in his mind with the trade across the Tiber;[12] and their presence among the inevitable "kissers" at Rome was obnoxious.[13] The choice of individuals whom he singles out of the laboring classes to receive opprobrium on this last score may be significant; for he selects the *textor, fullo, sutor;* that is, those who clothed him. Could it have been that the suggestion of unpaid bills was one of the things that made a wider berth desirable?[14] The phrase *avarus fullo*[15] in another passage carries suspicion with it. It is unfortunate that of the satirists only Martial offers testimony on fullers. Excavations at Pompeii give a more attractive impression of them. Two large fulleries and a smaller one which have been unearthed are suggestive of the extent of their business. *Graffiti* show that *fullones* themselves took an interest in town elections,[16] and their trade is one of those charmingly idealized in the graceful Cupid and Psyche pictures in the House of the Vettii.[17]

XV
INSTITORES

As a legal term in the third century, *institor* designated one "qui tabernae locove ad emendum vendendumve praeponitur, quique sine loco ad eundem actum praeponitur."[1] This definition from Paulus apparently involves two classes of men: namely clerks, agents, or managers in charge of any business concerned with buying and selling; and also street venders, hawkers, or peddlers. A study of the word as it occurs in a number of literary passages of the period under investigation supports the view that at this time the second signification was the one in common use.

Livy, for instance, in disclosing the parentage of C. Terentius Varro, who lived at the end of the third century B. C., says of his father: "Lanium fuisse . . . ipsum institorem mercis."[2] Although at this republican

---

[12] *Id.* 6.93.1-4. See pp. 18 f.
[13] Mart. 12.59. Cp. 7.95; 11.95, 98; Friedländer-Magnus 1.90-93.
[14] See pp. 75 f. for another explanation.
[15] Mart. 6.93.1.
[16] CIL. 4.998, 2966.
[17] Overbeck 390-396; Mau-Kelsey 335-336, 393-397.
[1] Dig. 14.3.18.
[2] Liv. 22.25.19. See pp. 28 f.

date it may have been considered degrading to manage one's own business,³ yet the historian's words may convey even a deeper cause for scorn. We know that it was customary in Martial's time for butchers to have their meat sold on the streets. "Omnia crudelis lanius per compita portat,"⁴ he tells us. The practice has persisted until our own day and was probably no new one in the first century of our era, so that it may be justifiable to read in Livy's statement a reference to the "butcher's boy" who delivered his master's wares. Horace couples *institor* with *navis magister* and *nautae*,⁵ thus combining "paddlers of the seas" with "peddlers of the streets." Ovid alludes to a vender of commodities which appealed to feminine interests with the distich:

> Institor ad dominam veniet discinctus emacem,
> expediet merces teque sedente suas,⁶

and Propertius says of a fishmonger:

> Suppetat hic, pisces calamo praedabor, et ibo
> mundus demissis institor in tunicis.⁷

Seneca complains that the extravagance of his day demanded numerous venders of iced water: "Habet institores aqua, et annonam, pro pudor! variam."⁸ The *popinarum institores*⁹ of the same author must be claimed as hawkers, not only because they are described by the phrase, "mercem sua quadam et insignita modulatione vendentes,"⁹ but because their counterpart is found in Martial's verses, "fumantia qui tomacla raucus / circumfert tepidis cocus popinis."¹⁰ Quintilian's metaphorical *institorem eloquentiae* is qualified by the expressive adjective *ambitiosum*.¹¹ Analogy would plausibly consign Juvenal's *institor hibernae tegetis niveique cadurci*¹² to the peddler class, even if reason for it were not at hand in Ulpian's note in the Digest: "Etiam eos institores dicendos placuit, quibus vestiarii, vel linteari dant vestem circumferendam et

---

³ Cp. Cic. *Flac.* 18: "Opifices et tabernarios atque illam omnem faecem civitatum quid est negoti concitare?" Cp. *Id. Dom.* 13.
⁴ Mart. 6.64.21.
⁵ Hor. *Epod.* 17.20; *Carm.* 3.6.30 f.
⁶ Ov. *Ars* 1.421 f.
⁷ Prop. 4.2.37 f.
⁸ Sen. *Nat.* 4.13.8. Snow and ice were packed in storage plants, *reponendae nivis officinas*, and preserved with straw; iced water was carted around by pack animals, *Ib.* §§ 3-9; cp. Petron. 31.
⁹ Sen. *Epist.* 56.2.
¹⁰ Mart. 1.41.9 f.
¹¹ Quint. 11.1.50; cp. 8.3.12.
¹² Juv. 7.221.

distrahendam, quos vulgo circitores appellamus."[13] The *sulphuratae lippus institor mercis* of Martial's lines[14] is of course a street vender, as is clear from the terms *ambulator* and *proxeneta* which the poet elsewhere applies to one who sold sulphur matches in exchange for broken glass.[15] Our last passage from Martial is one of special interest; it is the opening verse of the epigram which commends Domitian's law requiring shopkeepers to stay within bounds:

Abstulerat totam temerarius institor urbem.[16]

Now if *institor* here is merely a synonym for *tabernarius*, as most editors understand it, we shall have to infer that the word also signified shopkeeper at this period. But the references quoted above have rather implied that such a meaning was probably obsolete except in legal parlance. If this be true, Martial has again shown his adroitness in punning. The Roman reader's first concept from *institor* would be of a peddler; but as the epigram unfolded, he would grasp the poet's subtle purpose and no doubt laughingly agree with him that *tabernarii* who, like peddlers, carried their business out to the people were not really shopkeepers but street venders, and therefore well deserved the name *institores*.

Those whom the Romans classed as peddlers are naturally not accredited with flattering attributes. They were usually slaves[17] and are described as *discinctus*,[18] *demissis in tunicis*,[19] and are termed *viles pueri*,[20] and *vernae*.[21] Like many a salesman in the Latin countries of to-day, they set no *prix fixe* upon their goods, but took what they could get from the individual purchaser: Seneca calls attention to this trait in water-dealers;[22] Juvenal, in sellers of fabrics.[23] Venders whose wares gained them admission to the women of the household were bold, corrupt,

---

[13] Dig. 14.3.5.4; cp. Plin. *Nat.* 18. 225, where the setting of the constellation Vergiliae is called a harbinger of winter, like the peddler of heavy garments, *vestis institor*; but the text of the passage is unreliable.

[14] Mart. 12.57.14.

[15] *Id.* 1.41.3-5; 10.3.3 f. Cp. Stat. *Silv.* 1.6.73 f.; Juv. 5.48.

[16] Mart. 7.61.1. The clever suggestion has been made to me that *institor* here may refer to peddlers with push carts, but the general context seems to be opposed to this, especially verses 2, 5, 8, 10. See p. 61.

[17] They were sometimes freeborn, cp. Liv. 22.25.19; Dig. 14.3.1.

[18] Ov. *Ars* 1.421.

[19] Prop. 4.2.38.

[20] Mart. 1.41.8.

[21] *Id.* 1.41.2 ff.; cp. Pers. 4.21 f.; Mart. 10.3.1.

[22] Sen. *Nat.* 4.13.8.

[23] Juv. 7.220 f.

and immoral.[24] Hawkers were noisy and vulgar.[25] Horace mentions medical "quacks," *pharmocopolae*[26] in dissolute associations, and ranks the driver of a huckster's *caballus*[27] in the same category with a gladiator. To secure a contemptuous simile, Persius resorts to a comparison with a woman herbseller.[28] Martial describes the peddler of sulphur matches and broken glass as *lippus*[29] and *transtiberinus*,[30] links him with imposters and renegades, and includes in the same list venders of sausage and of pea soup[31] and the *viles pueri salariorum*.[32] Yet Petronius testifies that a man who had once been a peddler became a successful lawyer, and that another who used to carry wood on his back was able later to count his eight hundred thousand sesterces.[33]

The duties of the *pueri salariorum*, of whom mention has been made above, are matters for dispute. Marquardt denies that *salarii* were the same as *salsamentarii*, and maintains that they were retailers of salt.[34] Waltzing concedes that they may have been either "les mar-

[24] Hor. *Epod.* 17.20; *Carm.* 3.6.29-32; Ov. *Ars* 1.421 f.; *Rem.* 306; Juv. 6.591.

[25] Sen. *Epist.* 56.2; Petron. 68; Stat. *Silv.* 1.6.73 f.; Mart. 1.41.3 ff.; 10.3.1-4.

[26] Hor. *Sat.* 1.2.1-3; cp. Gell. 1.15.9 quoting Cato: "Auditis, non auscultatis, tamquam pharmacopolam. Nam eius verba audiuntur; verum se nemo committit, si aeger est."

[27]Hor. *Epist.* 1.18.36.

[28] Pers. 4.21 f. Cp. Petron. 7.

[29] This adjective in itself, of course, is not necessarily derogatory. Bad eyes seem to have been common among the Romans: Horace calls himself *lippus* (*Sat.* 1.5.30 f.; cp. 1.7.3; *Epist.* 1.1.29), and as Habinnas reminds us in Petronius 68, Venus herself was cross-eyed! But Horace would doubtless have objected to *lippus* as a characterizing epithet for himself; it often implies sensual excess, and even its figurative use is usually disparaging, cp. Hor. *Sat.* 1.1.120; 1.3.25; Pers. 1.79; 2.72; 5.77; Petron. 28; Mart. 6.39.11; 6.78.2; 7.20.12; 8.9.2; 8.59.2; 12.57.14; 12.59.9; Juv. 10.130.

[30] See pp. 19, 22.

[31] Some of these were sent out by cook-shops, but others seem to have been independent: Sen. *Epist.* 56.2; Mart. 1.41.6, 8, 9 f. Cp. Hor. *Epist.* 2.3.249; Petron. 14; Mart. 1.103.10. There was apparently some contrivance for keeping the food heated, cp. Juv. 3.249-253 and Darem.-Saglio 1.2.1502, fig. 1939. Seneca, *l.c.*, mentions hawkers of pastries and confections. See p. 16.

[32] Mart. 1.41; 10.3; 12.57.11-14.

[33] Petron. 38, 46.

[34] Marquardt 2.469.3. He claims that *salinator* meant originally "Salinenarbeiter"; *salarius*, "Salzverkäufer." Later, he thinks, the two became interchangeable, and he cites Arnob. 2.38, *salinatores* (="Salzverkäufer"), *bolonas, unguentarios, aurifices, aucupes*, and CIL. 6.1152, DIVO / CONSTANTINO / AVGVSTO / CORPVS / SALARIORVM (="Salinenpächter") / POSVERVNT.

chands de sel en détail ou les marchands de saumure."[35] There seems to be little evidence for positive proof, but the conjecture "salt dealers" is reasonable on the analogy of *Via Salaria*, "Salt Road," and (*argentum*) *salarium*, "salt money."[36] Martial's *salarii*, however, certainly savor of fish, especially those to whom allusion is made in an epigram addressed to the book which he is sending to Apollonaris for criticism. If you meet with his approval, says the poet,

> Nec scombris tunicas dabis molestas.
> si damnaverit, ad salariorum
> curras scrinia protinus licebit,
> inversa pueris arande charta.[37]

It is quite patent that the lines refer to the custom of using worthless manuscripts to wrap fish;[38] they further imply that the dealers' slaves scribbled accounts or memoranda on the back.[39] It appears, therefore, that in Martial's time *salarii* did deal in *scombri*; but this discovery surely need not interfere with considering them salt retailers; for who would be more appropriately engaged in pickling and salting? It may well be that *salarius* meant to the Roman what "salter" does to us, "one who makes, sells, or deals in salt; . . . one who salts meat or fish."[40] *Salsamentarius* may have been applied commonly to peddlers of salt meat or fish,[41] many of whom were doubtless *viles pueri salariorum*. We are told that Horace's father was a *salsamentarius* and that this was cast up to his son as a deep reproach; but the story, which is found in Suetonius's life of the poet, is probably an interpolation.[42] Hawkers

[35] Waltzing 2.226 f. He holds that *salinatores* were commercial speculatores, and he does not believe that the word *salarii* in CIL. 6.1152 designates those who were connected with salt mines.

[36] Cp. Plin. *Nat.* 31.89.

[37] Mart. 4.86.8-11.

[38] So also Pers. 1.43, "scombros metuentia carmina"; Mart. 3.2.4, "cordylas madida tegas papyro"; 3.50.9, "quod si non scombris scelerata poemata donas"; 13.1.1, "ne toga cordylis et paenula desit olivis." See p. 68.

[39] The context seems to demand this interpretation rather than to refer *pueris* to "school children," as most editors explain it. Why change to school children at the end, when the warning throughout has been against salt fish dealers? That the backs of old MSS. were utilized for accounts, the Oxyrhynchus Papyri give evidence. On the use of worthless MSS. for scribbling and trivial writing in general, cp. Horace *Epist.* 1.20.10-13 and *schol.*; Mart. 8.62.

[40] Century Dictionary.

[41] Wölfflin, *Archiv. Lat. Lex.*, 10, considers *salsamentarius* a dealer in salted foods of all kinds.

[42] Reifferscheid, ed. of Suet., 44 (Leipzig, 1860), cp. critical note.

of this kind were apparently a worthless type, yet from their dregs, it is said, there issued a *princeps equitum*[43] favored by Domitian, despised by the people, and designated by Juvenal, "pars Niliacae / plebis, . . . verna Canopi / Crispinus."[44]

XVI

LANII

Butchers may have been among those collected into the ninth general guild which Plutarch says was established by Numa.[1] Livy refers to a butcher-shop in the Forum in his story of Virginia which relates to the year 449 B. C.[2] Since, however, the historian may be guilty of an anachronism in this case, his words cannot be taken for conclusive proof of the presence of *lanii* in the Forum at the beginning of the republican period. The earliest epigraphical evidence for a *collegium laniorum* belongs apparently to about the middle of the second century B. C.; it records a corporation instituted near what was later the Piscina Publica.[3]

The butcher's trade appears to have been quite extensive and lucrative. Livy mentions a *lanius* of the third century B. C., who had acquired a large fortune.[4] By Cicero's time many meat-dealers, apparently, were required in order to administer to the extravagant tastes of high livers.[5] Trimalchio, in Petronius, selecting *laniones et unguentarii* to typify shopkeepers who sell by weight, maintains that they were born under *Libra*.[6] Martial considered a *lanius* to be essential even in an obscure country hamlet: "Give me a tavern," he says, "a butcher-shop, a place to bathe, a barber, draughtsmen and board, a few books personally selected, a congenial companion, a tall slave of enduring comeliness with a sweetheart to content him: give me these, Rufus, even at Butuntum, and you may keep for yourself the Thermae of Nero."[7] In another passage also Martial points to the butcher's omnipresence,

[43] Juv. 4.32 f.
[44] *Id.* 1.26-29; 4.28-33, 108 f.
[1] Wezel and Blümner are of this opinion. See pp. 1; 20, n. 2.
[2] Liv. 3.48.5.
[3] CIL. 6.167 f.; Waltzing 1.88.5; 4.26.65; Blümner, Müller's *Handbuch* 4.2.2. 593.6. The date is determined by the orthography: *AI* for *AE*, *EIS* for *I* (nom.), *OI* for *U*, *M* omitted; cp. Ritschl, *Opuscula Philologica*, 4.765 (Leipzig 1878); Egbert, *Latin Inscriptions*, 407 (New York, 1896).
[4] Liv. 22.25.18 ff.
[5] Cic. *Off* 1.150.
[6] Petron. 39.
[7] Mart. 2.48.

not only by the association in which he places him, but by the mere mention of him; for he singles him out together with the *tonsor, copo,* and *cocus* to represent the inconsiderate shopkeepers who had blockaded the sidewalks, until Domitian's law had demanded that they stay within bounds.[8] It is hardly to be supposed that the poet would have made special note of only a rare offender. There may be the further inference, too, that Martial was a forerunner of the advocates of sanitary methods, and therefore called attention here to those whose business was most unsuited to an open thoroughfare.

The unfeeling butcher, *crudelis lanius,* sinned against hygienic law, it seems, in still another direction; for although he found it advantageous to his trade and convenient for his patrons, to provide a prototype of the now familiar butcher's wagon and have his meats transported *omnia per compita,* he evidently did not deem it necessary to furnish a covering for his wares or to take the precaution of previous inspection as to savor and freshness![9]

Butchers in their social aspect have been maligned because of Livy's denunciation of C. Terentius Varro, of whom he renders the following report: "Loco non humili solum sed etiam sordido ortus. Patrem lanium fuisse ferunt, ipsum institorem mercis, filioque hoc ipso in servilia eius artis ministeria usum."[10] This is sometimes quoted to prove the ill repute not only of *lanii,* but of the industrial population in general.[11] There are three considerations, however, which strip the excerpt of much of its assumed implication. In the first place, any disparagement conveyed in the lines may express the attitude of the republican period only; for Livy may merely be copying the words of his source. Then too, no matter whose sentiments are expressed, it is to be noted that despite what may have been this man's due in theory, he managed, in reality, to win high position: he studied law, gained great influence, though by dishonorable cases, it was said, and actually ran the full gamut of the *cursus honorum,* becoming quaestor, plebeian and curule aedile, praetor, and finally consul and one of the commanding generals at the battle of Cannae, even receiving a magnificent ovation from senate and people upon his return from that disastrous struggle.[12] The story of his early life may not have been true, but it doubtless gained

[8] *Id.* 7.61. See p. 61.
[9] Mart. 6.64.18-21.
[10] Liv. 22.25.18 f.
[11] *E.g.*, Fowler 44; Kühn 10.
[12] Liv. 22.26; 56.1-3; 61.14; Plut. *Fab.* 18.

credence and therefore proves that as a matter of fact, lowly or even disreputable birth did not necessarily *per se* keep one from rising to power, even in the aristocratic days of the Republic. The third suggestion that presents itself is, that it would be no illogical interpretation of Livy's lines to balance *humili* with *lanium*, *sordido* with *institorem*, and refer *servilia* to the latter.[13] It would appear then, that Varro was of *humble* birth because he was the son of a butcher, but his heritage was *ignobile* as well; for his father had peddled his meats himself, performing the service which was usually consigned to a slave, *institor*;[14] he had even employed his son in these servile duties. Since, therefore, the heaviest aspersions in the passage seem to be directed against a slave's occupation and a lawyer's dishonesty, they cannot fairly be cited as committing *lanii* and other tradesmen to the depths of social disgrace and degradation. All that is clearly proved is that butchers were of humble station; and the references of the satirists add no more definite information.

XVII

Mangones

Men who earned their living by traffic in human lives seem not to have been able to purchase respect with their ill-gotten gains. Whether they were installed *ad Castoris* in shops which, according to Seneca, were crammed with a mob of slaves of the worst type;[1] or whether they had their platforms, *catastae*, concealed from the rabble's gaze in the secluded rooms of some respectable quarter such as the Saepta,[2] and there, to the inspection of the elect, displayed the "flower of Asia,"[3] which yielded from one to two hundred thousand sesterces per capita;[4] the word *mango* designated the class and branded them all "tricksters,"

[13] Otherwise the adjective is unnecessary. I consider that both *lanium* and *institorem* refer to *patrem*, *institorem* being rendered more emphatic by the intensive *ipsum*; *-que*, therefore, joins *fuisse* and *usum*.

[14] See pp. 22-27, especially pp. 22 f.

[1] Sen. *Dial*. 2.13.4.

[2] Mart. 9.59.1-6. See pp. 69 f.

[3] Juv. 5.56. Great numbers of slaves, many of them especially prized, came from the East, cp. Hor. *Epist*. 1.6.39; Petron. 31, 44, 63, 69; Mart. 7.80.9 f.; 10.76.3; Juv. 11.147; Wallon 2.48-50.

[4] Hor. *Epist*. 2.2.5 sets 8000 HS. as a conservative price for a slave. According to Petron. 68, a clever slave, a Jack-at-all-trades, was purchased for 300 denarii. Mart. tells of an undesirable girl who would not bring 600 HS. But cp. Mart. 1.58.1; 2.63.1; 3.62.1; 11.70.1; Juv. 5.56-61; Wallon 2.160-176; Marquardt 1.173-175.

if we are right in following the commonly accepted etymology from the Greek μάγγανον, μαγγανεύω.[5] The name was apparently no misnomer. They were adepts in magnifying the fine points of the slaves which they had for sale; but as Seneca charges: "Quicquid est, quod displiceat, aliquo lenocinio abscondunt, itaque ementibus ornamenta ipsa suspecta sunt: sive crus alligatum sive brachium aspiceres, nudari iuberes et ipsum tibi corpus ostendi."[6] The philosopher descries no trace of good in the cheap slave dealer, "sub quo nemo nisi malus est."[7] Martial declares *mangones* to be criminal in their greed.[8] Persius and Juvenal name their business as the last resort of the unprincipled,[9] and Quintilian characterizes a *mango* as a man cruel in rage, "qui non erubescit, nihil observat; etiam periculose avarus est."[10]

XVIII   XIX

MERCATORES   NEGOTIATORES

Roman trade was said to have received its first encouragement in the time of Numa.[1] After a gradual development in local fairs and markets, upon the opening of a grain trade with the city of Cumae, it appears to have expanded on the economic side into an interstate commerce; and in its religious phase, into the worship of Mercury,[2] a deity who, according to Professor J. B. Carter, was patterned after the Greek

---

[5] This is approved by Walde 461, and is sanctioned by most editors of the satirists who explain the word. Wilkins, however, in a note on Hor. *Epist.* 2.2.13, disagrees; he connects it with oE. *mangere* "a dealer," Germ. *-menger* from *mangian* "to traffic," ultimately *mang* "a mixture." It could be used of any dealer, it seems, who attempted to enhance the value of what he had to sell by clever mixing, furbishing, or misrepresentation; *e.g.*, a polisher of jewels (Plin. *Nat.* 37.199), a mixer of medicines or perfumes (*Ib.* 12.98; cp. 24.35), a dealer in wine (*Ib.* 23.39) or mules (Suet. *Vesp.* 4.3); but its specific signification in the Empire was evidently "slave trafficker," cp. Dig. 50.16.207: "Mercis appellatione homines non contineri, Mela ait: et ob eam rem mangones non mercatores, sed venaliciarios appellari ait." At all events it seems to imply regularly a deceitful dealer.

[6] Sen. *Epist.* 80.9. Cp. Hor. *Epist.* 2.2.3-15, with 11½vv. of virtues to 1½ of vice; Plin. *Nat.* 24.35; Quint. 2.15.25; Mart. 6.66.

[7] Sen. *Dial.* 2.13.4.

[8] Mart. 9.5(6).4 f.; cp. 6.29.1, *avarae catastae*.

[9] Pers. 6.77; Juv. 3.33, see pp. 46 ff.

[10] Quint. *Decl.* 340.

[1] Cic. *Rep.* 2.27.

[2] Hor. *Sat.* 2.3.25; Pers. 5.112; Petron. 29, 77.

Hermes Empolaios, the protector of merchants.³ Although this divinity later figured in Latin poetry with many attributes and functions of Hermes, Professor Carter believes that in actual cult "he never regained the many-sidedness which he had lost in coming to them (the Romans) merely as a god of trade."⁴ A temple was built to him near the Circus Maximus, and upon its dedication in 495 B. C. there was instituted a *collegium mercatorum*,⁵ an embryonic Chamber of Commerce, whose members were known as *mercuriales*.⁶

During the Republic Rome's commercial activities steadily increased and became so important that Mommsen⁷ has put forth the theory that her territorial expansion at this time was due to mercantile influence. Professor Tenny Frank, in his volume on *Roman Imperialism*,⁸ opposes this view on the ground that the evidence offered does not bear examination; he concludes that traders did not exert any considerable influence upon the policies of the senate. His refutation is ingenious though not thoroughly convincing, especially in view of such a statement as Cicero's: "Maiores nostri saepe mercatoribus aut naviculariis nostris iniuriosius tractatis bella gesserunt."⁹ Perhaps a middle course between the two extreme opinions would more nearly approach the truth.

In the first century of the Empire, trade was so extensive that Claudius, carrying out a project contemplated by Julius Caesar, made vast improvements at Ostia to provide a safe and convenient harbor.¹⁰ With the completion of this great work, a fresh impetus must have been given to mercantile interests, which under the emperors became worldwide in scope. From random references in the satirists, we learn that caravans brought pepper, gold, ebony, and ivory from farthest India;¹¹

---

[3] Carter 77 f.

[4] *Id*. 79.

[5] Liv. 2.27.5.

[6] Cic. *Ad. Q. Fr.* 2.5.2; cp. Wissowa 5.4.304 f.

[7] Mommsen-Dickson, *Hist. of Rome*, 3.238 f., 274, 295, 415 f., 421; cp. Ferrero, *Greatness and Decline of Rome*, 38 (New York, 1907).

[8] Ch. 14, "Commercialism and Expansion." See p. viii.

[9] Cic. *Manil*. 5.11.

[10] Juv. 12.75-81 and *schol*.; Suet. *Claud*. 20.1; Dio 60.11; Lanciani, *Ann. dell Inst*. 40 (1868). 144 ff.

[11] Hor. *Carm*. 1.31.6; *Epist*. 1.1.45; Pers. 5.135 f., cp. 54 f. and Virg. *Georg*. 2.116 f.

China furnished silk;[12] Arabia, incense;[13] Phoenicia, dyes;[14] Syria, spices and perfumes;[15] Cilicia, saffron.[16] Voyages were made to Phrygia and Bithynia for timber and marble,[17] to Pontus for herring, oil, and tow.[18] Crete and the islands of the Aegean[19] were active in commerce. But the *vagus mercator*[20] did not stop with the East; leaving the Aegean, he made for the African Seas and passed between the Pillars of Hercules, far out into the Atlantic.[21] Every quarter of the world[22] from the rising to the setting sun[23] delivered its own special products.

It was not without great anxiety of spirit and peril of life[24] that these distant journeys were made. The discomforts on shipboard were many. Even the captain of the enterprise might have to eat his dinner from the rowers' bench, propping himself against a coil of rope and partaking of wine that would probably be flat.[25] His vessel was at the mercy of wind and storm, and partial or total shipwreck was not uncommon.[26] Indeed *dis carus ipsis*, as Horace maintains, was he who sailed to the Atlantic three and four times a year and came back safe.[27]

Yet the trader's calling seems to have had no dearth of devotees: "Behold harbor and sea teeming with big vessels," says Juvenal, "the greater portion of mankind is now on the deep."[28] Was it the danger

---

[12] Petron. 119; Mart. 11.27.11.

[13] Pers. 5.135, cp. Virg. *Georg.* 2.117.

[14] Hor. *Sat.* 2.4.84; *Epist.* 1.6.18; Mart. 8.48. 1; 9.22.13; 10.87.10; *et passim;* Juv. 1.27; 7.134.

[15] Hor. *Carm.* 1.31.12 and *schol.*; 2.7.8.

[16] *Id. Sat.* 2.4.68; Mart. 3.65.2; 9.38.5; Juv. 14.267 and Mayor's note.

[17] Hor. *Carm.* 1.35.7 f.; 3.1.41; *Epist.* 1.6.33; Mart. 9.75.7 f.

[18] Pers. 5.134 f.

[19] Pers. 5.135; Mart. 1.88.3; Juv. 14.270 f.

[20] Hor. *Epist.* 2.3.117.

[21] *Id. Carm.* 1.31.11-14; Juv. 14.278-280.

[22] Pers. 6.76.

[23] Hor. *Sat.* 1.4.29 f. For a highly colored description of the extent of Rome's commerce at the end of the Republic, cp. Petron. 119.

[24] Hor. *Epist.* 1.1.44-46.

[25] Pers. 5.146-148.

[26] Hor. *Carm.* 1.1.13-17; 3.1.26; *Sat.* 1.1.6; Pers. 6.27-31; Petron. 76, 114; Mart. 4.66.14; 5.42.6; Juv.12.17-82; 14.268, 292-302.

[27] Hor. *Carm.* 1.31.13-15.

[28] Juv. 14.275-277.

and excitement that lured them; or was it the air of importance with which they could fairly bristle upon their return, when as *tumidi negotiatores*[29] they would be able to play the rôle of Othello and prate

> Of antres vast and deserts idle,
> . . . .
> And of the Cannibals that each other eat,
> The Anthropophagi, and men whose heads
> Do grow beneath their shoulders?[30]

All this made its appeal perhaps, but in the eyes of the satirists the real inspiration came from avarice. She it is, we gather from Horace, who drives men indefatigably over every sea and headlong into danger. They strive to be first in port and to lose no bargain; for they must procure the round sum of a thousand talents, then another thousand, and another, and still another, until they have made their fortunes square. In the midst of a storm they may long for the ease of home, but they later repair their battered ships, *indociles pauperiem pati*.[31] Avarice makes her demands, Persius adds, even upon the man who loves the luxury and comfort of home. Rousing him from his snoring slumbers, she goads him on to load his slave with packing-skin and wine-jug and to set out for the hardships of a voyage, merely that he may be able to squeeze a greedy eleven percent from the money that he had been nursing at Rome for a modest five. She would have him sell his very soul for gain.[32] Petronius, too, notes the high interest that buoys up the man who trusts the sea.[33]

---

[29] Mart. 10.87.9, cp. 104.16. On the history of the change in meaning of *negotiator*, cp. Ernesti's treatise, *De Negotiatoribus Romanis*. By abundant evidence he proves quite satisfactorily that throughout the republican period, the *negotiator* corresponded in the provinces to the *faenerator* at Rome; it became part of his business, however, to attend to the shipment of the state grain supplies, and since to this extent his duties coincided with those of the grain merchant, the terms for the two became confused, and in the Empire *negotiator* was sometimes used interchangeably with *mercator*. See p. 36. Martial's *tumidus negotiator* was evidently an importer with a shop in the Portico of Agrippa, cp. Juv. 6. 153-157. Cp. Petron. 43, *homo negotians*, referring to a wine merchant; *Id*. 76, *negotiari*, used in connection with wine and other merchandize; *Id*. 116, *negotiatio, negotiatores*, designating broadly business in general. On Mart. 11.66.2 where the original meaning of *negotiator* seems to be retained, see pp. 36 f.

[30] Shakesp., *Othello*, 1.3.140-145; cp. Juv. 12.81 f.; 14.281-283.

[31] Hor. *Carm*. 1.1.13-18; 1.31.10-15; *Sat*. 1.4.25 f., 29-32; *Epist*. 1.1.42-48; 1.6.31-38.

[32] Pers. 5.52-55, 111 f., 132-150; 6.75-80.

[33] Petron. 83.

Juvenal is the bitterest critic. In one vivid, caustic description he sums up the whole situation: the broad extent, the danger, and the ultimate aim of foreign trade. The passage is familiar, but its interest makes it worth recalling. It is better than any play, the poet proclaims, to watch the peril that the avaricious man incurs in his struggle for wealth. Do tight rope walkers or acrobats hurled from the springboard furnish the mind more delight than you who tarry forever in your Corycian bark, offering a life-long plaything to Corus, the northwester, and Auster's southern blasts—you reckless and worthless merchant of an odorous bag of merchandize?[34] . . . A fleet will go wherever prospects of gain shall call. . . . Yet the grand reward of your exertion is that you may return home from your voyage with bulging purse, proud of your swollen money-bags, and may boast that you have seen ocean monsters and the young folks of the sea. . . . Though he does not tear his tunic and cloak, that man is in need of a keeper, who fills his ship even to the very bulwarks with cargo and is separated from the waves only by a plank, since the incentive to this great hardship and risk is but a silver coin stamped with the Emperor's miniature and superscription. A tempest threatens, yet "Cast off the hawsers," cries the owner of the grain and pepper that he has bought up, "this colored sky, this strip of black cloud portend no ill; it's only a bit of heat lightning." Poor wretch, perhaps that very night his ship will be shattered; overboard he'll go, and all but sink overwhelmed by the billows, as he grasps his money-belt with his left hand and his teeth. Then he whose desires were not sated a while ago with all the gold that the Tagus and the Pactolus roll in their ruddy sands will now be satisfied with a morsel of food, and a few rags to cover his shivering loins, while as a shipwrecked mariner he asks for a penny and maintains himself by his painted picture of a storm at sea.[35]

Wide indeed was the merchant's range and hazardous his lot. Juvenal's account, however, is certainly much exaggerated. It is hardly probable that every quiet sea returned a millionaire to the moneyed aristocracy, and that every angry one was destined to set a mendicant adrift in the dregs of the populace. Of those who begged at Rome, carrying a picture of their disaster at sea painted on a fragment of the vessel, some were probably imposters;[36] others may have been the surviving *nautae* of a shipwrecked crew, and would, therefore, come for

[34] See p. 38.
[35] A paraphrase of Juv. 14.256-302.
[36] Pers. 1.88-91. Cp. *Id*. 6.27-33; Petron. 115; Mart. 12.57.12.

the most part from the ranks of slaves.[37] But since the promotors and managers of mercantile projects not only required strong financial backing, but also gained great profits, we may with reason suppose that they, in the majority of cases, were members of the *equites*;[38] or that in the course of events they became citizens with equestrian rating; indeed, Trimalchio informed his guests that he had made ten million sesterces on a single voyage.[39] Some men of means were apparently interested in merchant ships merely as an investment;[40] others were doubtless capitalists who financed undertakings but had the actual business carried on by freedmen.[41] On this point, Professor Frank, speaking for the republican period, expresses the belief that "traders in the provinces were looked upon at home as a somewhat low class of adventurers, who had little connection with the vital interests of the state," and that wealthy Roman citizens, even though they had made their money in foreign commerce, yet being "always lovers of terra firma, gradually drifted into capitalistic enterprises on land, leaving the freedmen of Oriental and Greek stock in Italy and their sons to gain control of the shipping."[42]

Although it is possible that the last part of this statement may hold for the early Empire, too, yet if the evidence from the satirists is at all reliable, we must suppose that many freedmen also engaged in maritime trade. Juvenal's assertion, "Plus hominum est iam / in pelago,"[43] speaks for the popularity of commercial pursuits; both Horace and Persius refer to the competition that prevailed,[44] and the long satirical passages in Persius and Juvenal upon the merchant's life are presumably directed against men of the higher classes.[45] Considering the

[37] Cp. Hor. *Sat.* 1.5.11, 16, 19, *nautae*, of canal boatmen on a tow path; Mart. 10.85.1, on a *nauta* of the Tiber; Juv. 8.174.
[38] Friedländer-Magnus 1.143; Fowler 26; Tucker 238.
[39] Petron. 76, cp. 43.
[40] Cp. Petron. 141; Mart. 4.66.14; 5.42.6.
[41] Cp. Petron. 76.
[42] Frank, *Roman Imperialism*, 286, 289. Cp. Frank's interesting observations in an article on "Race Mixture in the Roman Empire," Amer. Hist. Rev. 21 (1915-1916). 689-708; he concludes from inscriptional evidence that in the time of Juv. and Tac., probably 90% even of the *free* plebeians had Oriental blood in their veins, and that in the melting-pot of the whole Empire, the Oriental formed a very large part of the amalgum.
[43] Juv. 14.276 f. Frank, in the article last cited, at one time (695) discounts similar "sweeping statements" of Juv., at another (690) takes them at their face value.
[44] Hor. *Epist.* 1.6.32 f.; Pers. 5.136.
[45] Pers. 5.52 ff., 132 ff.; Juv. 14.265 ff.

greed for amassing fortunes in the first century after Christ, and the possibilities that foreign trade held for satisfying it, it is altogether likely that some financial corporations gave their attention solely to this line of gain and for greater security managed their own enterprises; in such cases capitalist and merchant would be one and the same. This would furnish another cause for the practice which arose in the imperial period of using the terms *negotiator* and *mercator* interchangeably; not only did the individual designated by the former (that is, according to Ernesti's theory, the money lenders in the provinces) enter the latter's domain by superintending the grain trade, but the merchant, by uniting with others of his kind into stock companies for business on a large scale, assumed to a degree the original nature of the *negotiator*.[46] On the other hand there must have been a tempting inducement for independent local tradesmen who were shrewd and successful, to broaden their interests and do their own importing; in this way freemen of the lower ranks, as well as freedmen, would enter the lists of *mercatores*.

Negotiatores

Public opinion seems to have varied at different periods in regard to those engaged in commercial activities. The above characterization of them as a "low class of adventurers" would no doubt have received Cicero's sanction; for in accounting for the Carthaginians' proneness to lie and cheat, he laid it to their natural location on a good harbor, which had caused them to be associated constantly with merchants and foreigners, and had therefore enticed them to the pursuit of deceit in their eagerness for gain.[47] Again, it is with biting irony and malicious intent that he says of Verres: "Mercatorem in provinciam cum imperio ac securibus misimus."[48] But even Cicero leaves a loop-hole. While branding small trade as degrading, he adds that it is not only not reprehensible but is even laudable for a person to engage in extensive mercantile enterprises, especially if, after having made reasonable profits, he shall retire straightway to landed estates.[49] One would be inclined to believe that the philosopher himself had interests in *mercatura . . . magna et copiosa!*

Among the references to traders, from which we may judge of the attitude toward them under the Empire, we find one in which Martial accuses a disreputable character of being a *delator, calumniator, fraudator*,

[46] See n. 29.
[47] Cic. *Leg. Agr.* 2.95; cp. *Off.* 150.
[48] *Id. Verr.* 4.8.
[49] *Id. Off.* 1.151.

negotiator, and other despicable things.⁵⁰ Fortunately for the merchant, it cannot be proved that *negotiator* is here used synonymously with *mercator*, as it sometimes was during the Empire.⁵¹ On the contrary, it is more probable that the word, which in this case is apparently equivalent to "swindler," has its common Ciceronian meaning of *faenerator*;⁵² the inference is supported by the closing words of the epigram: "Miror / quare non habeas, Vacerra, nummos." Petronius hints that a *homo negotians* will never do well, unless he is suspicious and distrustful of others, and is therefore (by inference) inclined to dishonesty himself; yet the wine merchant whom this very passage lauds for his liberality and his trusting nature, is said to have succeeded phenomenally.⁵³

Some of the citations from the satirists which have been noted on the preceding pages sound quite derogatory, but they can scarcely be considered a safe criterion for obtaining a sane judgment of the general attitude of the time to which they refer; for practically all of them are directed against excessive attention to trade, which was resulting in a struggle for wealth for money's sake only. Persius, the philosophic recluse, who probably knew the least about actual conditions in the business world, is especially bitter in his attacks. Striving to maintain in his fifth Satire the Stoic paradox that none but the philosopher is truly free, he argues that he who is under the influence of some overwhelming passion can offer no claim as a free agent. *Mercatura*, therefore, which to him symbolizes *avaritia*, he condemns as a sort of preliminary vice in a *cursus dedecorum* consisting of *luxuria, amor, ambitio*, and *superstitio*.⁵⁴ He also puts the merchant, named metaphorically for the avaricious man, into another group with doubtful associates, namely, the bonvivant, athlete, gambler, and debaucheé.⁵⁵ As shown in the sixth Satire, too, his animosity to trade seems due to the fact that he thought it as impossible to discover a man of moderation engaged in it as to find one who could answer Chrysippus' question as to when the pile has become a heap. The *mercator* whom he scores is the one who, to gain his ends, would stoop to dishonesty, or to slave-dealing,⁵⁶ the lowest form of traffic.⁵⁷

⁵⁰ Mart. 11.66.
⁵¹ Cp. *Id*. 10.87.9 f.
⁵² See n. 29.
⁵³ Petron. 43.
⁵⁴ Pers. 5.132-188.
⁵⁵ *Id*. 5.52-61.
⁵⁶ *Id*. 5.137; 6.75-78.
⁵⁷ See pp. 29 f.

It is apparently this same type that Juvenal had in mind in his fourteenth Satire; such was his *perditus ac vilis sacci mercator olentis*;[58] nowhere else is he so scornful, and here the text is corrupt; the Bücheler-Leo restoration *assiculis* lessens very much the sting of *ac vilis*. Furthermore Juvenal manifests a sincere affection for a certain Catullus[59] who was presumably a merchant[60] and had barely escaped shipwreck in a storm at sea. In gratitude for his friend's preservation, the poet makes sacrifice to his household gods and to Jupiter Capitolinus. On the day of the festivities, he writes to a mutual acquaintance that the occasion on which he is permitted to pay this honor is dearer to him than his own birthday; his motives for making it are entirely unselfish, and his only regret is that his offering cannot be more liberal.[61] His further declaration that there were none of Catullus's confrères so little addicted to avarice as Catullus himself,[62] may be ascribed to a biased judgment.

The following lines from the same hand are more noncommittal:

> Mense quidem brumae, quo iam mercator Iaso
> clausus et armatis opstat casa candida nautis,
> grandia tolluntur crystallina.[63]

Of them Duff in his edition says: "He (Jason) is called *mercator* sarcastically, because of the purpose of his voyage; the Argonautae are degraded to *nautae*." Certainly there is no positive indication in the verses themselves that this was Juvenal's intention, and it is a question whether he was wasting any irony upon the male sex at this point. Was he not rather venting all his sarcasm on extravagant women who demanded rich vessels of crystal from the expensive shops of the *Porticus Argonautarum* even at the time of the Saturnalia, when the Portico was possibly hidden from view by the canvas booths erected for the *sigillaria*,[64] the image-fair, to which most people were repairing to purchase ordinary figures of clay? The expedition of Jason and his comrades was of course an appropriate theme for mural decoration in a shopping district and its symbolism must have been evident to all, so that unless these heroes

[58] Juv. 14.269 (Jahn, 1851), the reading commonly adopted.
[59] *Id*. 12.1-30, 83-98.
[60] *Id*. 12.37-47, describing a rich cargo that had to be thrown overboard.
[61] *Id*. 12.1-16, 83-98.
[62] *Id*. 12.48 f.
[63] *Id*. 6.153-155.
[64] Suet. *Claud*. 5.1; *Schol*. on Juv. 6.154. Cp. Mart. 14.182; Suet. *Claud*. 16.4; *Nero* 28.2; Gell. 2.3.5; 5.4.1; Marquardt, *Staatsverwaltung*, 3. 563; Darem.-Saglio 4.2.1302; Smith, *Dict. of Antiq.*, 2.600 f.

had been painted on the Portico for their eternal disgrace and degradation, it seems hardly necessary to read into the lines just quoted any such implication as Duff suggests.

There is still another passage from Juvenal which displays no special ill-feeling against the *mercator*. In the seventh Satire, after deploring the lack of patronage granted to poets in his day, he hints that they had better find some other honorable vocation, and suggests *mercatura* together with *militia* and *agricultura* as suitable occupations.[65] Horace, too, grouped together in his lines the merchant, the soldier, and the tiller of the soil, and he seems to have been of the opinion that in the making of money any one of them had as great an advantage as another.[66] Indeed in the social discontent attendant upon the changing conditions of his time, he found them each envying the other, yet he felt assured that none of them would really wish to change his lot if he should have the opportunity.[67] And suppose merchants do amass a fortune, he reasons in one of his Epistles, to be sure they are slaves to their own desires and are deserters from the side of Virtue, but they are also useful to society; let them relieve the market and fill your larders and granaries, don't crush them.[68]

Because of this utilitarian value, if for no more generous reason, practical people like the Romans must, at all periods of their history, have recognized the importance of *mercatores* both socially and economically. From the very beginning of the Empire certainly, due to an example set by the Emperor himself, those engaged in mercantile pursuits seem to have met with much encouragement. In the ode addressed to Augustus as the savior of the state and society, Horace deems it not unfitting to invoke him in the name of Mercury, the promoter and patron of trade and commerce; and for him even in this capacity, he offers the fervent prayer:

> Serus in caelum redeas diuque
> laetus intersis populo Quirini.[69]

[65] Juv. 7.32 f. Cp. Petron. 116 where strangers approaching Croton are advised, if they be *negotiatores* (i. e., presumably, business men of any sort, see n. 29), to change their occupation; but if they are clever liars, they may hasten on to wealth: the town has the reputation of supporting two classes of people, legacy-hunters and their victims. See pp. 43 f., 48 f.
[66] Hor. *Sat.* 1.1.4-32; cp. *Epist.* 1.16.70-72; 2.3.117; Petron. 83.
[67] Hor. *Sat.* 1.1.1-19; cp. *Carm.* 1.1.11-18.
[68] *Id. Epist.* 1.16.67-72.
[69] *Id. Carm.* 1.2.45 f., cp. 41 ff.

XX

Pistores

The baker's trade was not one of long standing at Rome, and consequently was not represented among the early industrial colleges which Plutarch ascribes to Numa.[1] The omniscient elder Pliny explains the situation quite fully in his *Naturalis Historia*. There were no bakers at Rome, he says, until the time of the war with Perseus, more than five hundred and eighty years after the founding of the city; before that, bread had been made at home under the supervision of the women.[2] In the light of Pliny's further statement, we must interpret *pistores* in Plautus as "millers"; for he claims, on the authority of Ateius Capito: "Cocos tum panem lautioribus coquere solitos pistoresque tantum eos qui far pisebant nominatos."[3]

Later, however, when baking was introduced as a trade, in accordance with Greek practice apparently, it was adopted as an additional occupation by those who ground the grain; and miller and baker became identical, both designated by the common title, *pistor*.[4] Hence it is that, to quote Mau-Kelsey, "we rarely find in Pompeii—and then only in private houses—an oven without mills under the same roof."[5]

Although baking did not cease, of course, to be a home employment, and although the wealthy often maintained special *pistores* among their slaves,[6] bakeries, which were called *pistrina (ae)* or *furnariae*[7] from both the old and the new business conducted in them, received ample patronage. More than twenty of them have already been unearthed in Pompeii.[8] At Rome, according to the fourth century regionary catalogue, there were at that time[9] from fifteen to twenty-four in each *regio*.

Martial complains of the noise of the industry, which he declares to have been insufferable even before daybreak; in his own words:

---

[1] See p. 1.

[2] Plin. *Nat.* 18.107.

[3] *Ib.* 108; cp. Varro in Non. 223; Fest. 58 M.

[4] Cp. Mart. 8.16.4 f.

[5] Mau-Kelsey 388.

[6] Varro in Gell. 15.19.2; Petron. 38, 60; Mart. 11.31.8-10; 13.10. At times, no doubt, as had been the custom before the introduction of professional bakers (see n. 2; cp. Harcum 74 f.), *coci* were also charged with the baking, cp. Petron. 68.

[7] Sen. *Epist.* 90.22; Petron. 73; Plin. *Nat.* 7.135; 18.86; Suet. *Aug.* 4.2; *Vitel.* 2.1. For a discussion of bakeries, cp. Blümner, *Tech.*, 1.89-95.

[8] Mau-Kelsey 388.

[9] *Notitia* (cp. Jordan 2.541-564).

"Negant vitam . . . nocte pistores."[10] One score against them seems to have been the disturbing jargon of the hawkers, who were sent out, especially by the makers of pastries and confections, to sell their wares upon the street. Each had his own individual singsong cry and might begin to rend the air as early as cockcrow, in order to catch the small boys who were on their way to school and had been obliged to leave home too early for breakfast.[11] But a far more serious cause for grievance must have been the grinding and grating of the heavy mills. These were usually turned by asses and mules.[12] The use of horses for this purpose is mentioned by Juvenal in a verse which requires further comment because of a disputed reading. In giving some sound advice to reprobate nobles, whom he urges to live on their own honors instead of on the laurels of their forebears, he instances the case of steeds of excellent breed, which win no glory from a famous pedigree, if they are themselves *segnipedes dignique molam versare nepotes*.[13] Jahn, in his edition of 1851, and Mayor read *Nepotis* and presumably interpret it as referring to Nepos, a miller-baker. The form as quoted, however, seems to be correct beyond a doubt, both because it is found in the first hand of the best manuscript,[14] and because it accords better with the context.[15] If *nepotes* (v. 67) is considered as merely a repetition of *posteritas* (v. 62) and a term in contrast with *maiorum* (v. 64), a strong and desirable antithesis is obtained between the renown of noted ancestors and the ignominy of their unworthy progeny. Especially interesting and convincing on this point is the following inscription, which records the victories—and the lineage —of a famous race horse: HIRPINVS. N(EPOS). AQVI / LONIS . VICIT. CXIIII / SECVNDAS. TVLIT / LVI. TERT. TVL. / XXXVI.[16]

[10] Mart. 12.57.4 f.
[11] Sen. *Epist.* 56.2; Mart. 14.223.
[12] Ov. *Ars* 3.290; *Fast.* 6.311 f. For illustrations and a description of the whole subject, cp. Mau-Kelsey 388-392; Blümner, *Tech.*, 1.20-49.
[13] Juv. 8.67.
[14] Cp. Jahn-Bücheler-Leo edition, critical note and *Praef.* v ff., xiii, xv, xxii.
[15] The verses from 62 are:
  Sed venale pecus Coryphaei posteritas et
  Hirpini, si rara iugo victoria sedit;
  nil ibi maiorum respectus, gratia nulla
65 umbrarum; dominos pretiis mutare iubentur
  exiguis, trito ducunt epiraedia collo
  segnipedes dignique molam versare nepotes.
[16] CIL. 6.10069. Cp. Mart. 3.63.12; Friedländer-Freese 2.21-33; Friedländer-Gough 4.148-166.

Since the turning of a mill was such inglorious work even for beasts of burden, it would seem to have been no proper task for human beings, but they were sometimes forced to perform it. Slaves and criminals, for instance, might be sent to *pistrina* to suffer the penalty of hard labor in chains.[17] Naturally the work was very degrading and to employ freemen at it against their will, says Blümner, was strictly forbidden.[18] Mau-Kelsey point out that at Pompeii there were a number of small bakeries rather than a few large establishments;[19] this fact, coupled with literary and epigraphic evidence, leads to the conclusion that *pistores* were wont to specialize in trade along certain lines. Bakers of two varieties of bread mentioned by Pliny[20] are recalled in a particularly vivid manner in inscriptions. One reveals the *clibinarii* of Pompeii supporting a certain Trebius for the aedileship;[21] the other is on a tombstone dedicated by a wife to her beloved husband, M. Junius Pudens, a wholesale baker of Parthian (?) bread, CVM.QVO.VIXIT. A. VIRGINITATE. ANNIS . XXXV / SINE.VLLO.DOLORE.NISI.DIEM.MORTIS.EIVS.[22] Other inscriptions record a *pistor candidarius*,[23] a *corpus pistorum magnariorum et castrensariorum*,[24] a *corpus pistorum siliginiariorum*.[25] Among pastry confectioners[26] there were *crustularii*,[27] and *dulciarii*. One of the last

[17] Plaut. *Persa* 21 f.; *Poen*. 827 f.; Ter. *Andr*. 199 f.; *Phorm*. 249; *Wallon* 2.227.

[18] Blümner, *Tech.*, 1.33.3.

[19] Mau-Kelsey 388.

[20] Plin. *Nat*. 18.105 f.: oysterbread, *ostrearius*; cake bread, *artolaganus*; hurry bread, *speusticus*; oven bread, *furnaceus*; tin bread, *artopticeus*; mold bread, *in clibanis*; Parthian or water bread, *Parthicus, aquaticus*; Picentine bread, *Picentinus* (cp. Mart. 13.47). Petron. 66 speaks of whole wheat bread, *autopyrus*; for other varieties see Blümner, *Tech.*, 1.77-89.

[21] CIL. 4.677.

[22] CIL. 6.9810. The designation of the trade is in the line PISTORI.MAGNARIO. PEPSIANO. Editors before Mommsen understood the last word as PERSIANO=PARTHICO. He believes this wrong and thinks that the form found in the inscription may be connected with the Gr. πέψις = Lt. *coctura*; he refers to the breads which Plin. (*l.c.*) says were named *a coquendi ratione*. Blümner, *Tech.*, 1.92.8, suggests *Gesundheitsbrot*; compare our "Holsum Bread." The old explanation, however, seems as simple and reasonable as any; the misspelling is unimportant, cp. LIBERTARBVS in the last line.

[23] CIL. 14.2302; cp. Petron. 66.

[24] CIL. 6.1739.

[25] CIL. 6.22, cp. Waltzing 2.80. Cp. CIL. 6.1958.

[26] Cp. Blümner, *Tech.*, 1.94 f.

[27] Sen. *Epist*. 56.2; *libarii* here is the conjecture of Caelius Rhodiginus for *biberari* and *liberari(i)* of the MSS., cp. CIL. 4.1768.

is praised by Martial for his honey knick-knacks; to quote the poet:

> Mille tibi dulces operum manus ista figuras
> extruet: huic uni parca laborat apis.[28]

Bakers appear to have been especially prominent in the business world during the Empire. What reputation they enjoyed is a matter for conjecture. Suetonius tells us that Antony and Cassius of Parma taunted Augustus with being descended from a baker of Aricia.[29] Their scorn was probably characteristic of the prevailing attitude toward trade at the close of the Republic; yet it may be remarked that the biographer's anecdote, whether true or false (for a fictitious tale to be worth its fabrication requires a foundation or semblance of truth), shows what possibilities might be in waiting for a baker's scion.

The emperors doubtless encouraged *pistores*, in order to facilitate the distribution of flour and bread. From the time of the Republic, Waltzing notes, the aediles entered upon contracts with them, to enable the people to buy bread of good quality at a moderate price.[30] Martial, writing in Domitian's time, mentions the bakery with the wine shop as the natural place to spend one's last denarius.[31] Speaking of Trajan's reign, Aurelius Victor reports: "Annonae perpetuae mire consultum, reperto firmatoque pistorum collegio."[32] Since *reperto* and *firmato* have been considered contradictory terms, the emendations *recepto* and *reparato* have been suggested, and the explanation is offered that the guild was probably established earlier, but that Trajan gave it special privileges and settled its relation to the grain supply.[33] It may have been the strong incentives put before *pistores* at this time that caused Juvenal to choose their trade as representative of the money-making occupations which, he asserts, even famous and illustrious poets had been on the point of entering, had not a worthy patron of literature appeared in the person of the Emperor.[34] If there is this connection between the

---

[28] Mart. 14.222, cp. 223. For various confections of *pistores*, cp. Petron. 60,66; Mart. 11.31.8-10.

[29] Suet. *Aug.* 2.3; 4.2. It was said also that Vitellius's great-grandfather had married the daughter of a baker, cp. Suet. *Vitel.* 2.1.

[30] Waltzing 2.79; cp. Petron. 44.

[31] Mart. 2.51.1-3.

[32] Aur. Vict. *Caes.* 13.5. The corporation was under the supervision of the *praefectus annonae*. Waltzing 2.82 observes that such a *corpus* would become indispensable to the state in the third century, when, between the time of Alexander Severus and Aurelian, bread instead of flour was distributed free.

[33] Waltzing 2.79 and n. 5; cp. 4.37-39.

[34] Juv. 7.1 ff. See pp. 48 f.

poet's verses and a historic fact, the *Caesar* addressed in the Satire is undoubtedly Hadrian, and the disputed question on this point is settled.[35]

Although Juvenal's reference to the baker's trade in the passage just cited is doubtless ironical, nevertheless he maintains that it was at least an honorable means of livelihood, befitting free men, and far better than perjury or the practices of the *delator*, by which slaves, freedmen, and others were rising to influence and power. Two lines from Martial depict a *pistor* as a man of low character and immoral habits, but they refer to a slave, and presumably to one in domestic service.[36] The epigrammatist cannot resist a jibe, either, at the baker who became a lawyer and was trying to make two hundred thousand sesterces. As fast as he made money, however, he squandered it. This characteristic Martial describes in a jesting metaphor which clearly gives his view on the mooted question of the "leopard's spots"; it reads:

A pistore, Cypere, non recedis:
et panem facis et facis farinam.[37]

Bakers who did not indulge in the impulse to change their occupation, as Cyperus did, were by no means doomed therefore to suffer financial straits. Archaeology bears witness to this fact; for the tomb of M. Vergilius Eurysaces at Rome near the Porta Maggiore, which is wrought with reliefs portraying the processes of the baking business, proves by its elaborate proportions and minuteness of detail[38] that it sheltered the last remains of one who in life was a captain of industry and who felt that neither shame nor degradation was attached to his trade.

XXI

Praecones

The term *praeco* appears to have been applied by the Romans to several classes of men whose duties were quite distinct but who, as Cicero expressed it, "employed their voice as a means for gain."[1] Juvenal, perhaps ironically, uses it once instead of *nomenclator*[2] for the domestic slave who knew all his master's clients and whispered their

---

[35] The supposition that Hadrian is meant is now commonly accepted by editors, cp. also Friedländer-Gough 4.312-315. Hermann, K. F., *De Juvenalis Satirae septimae temporibus* (1843) and Teuffel-Schwabe-Warr 2.§ 330.2 argue for Trajan (but cp. *Id*. § 331.4), and Nettleship, Jour. of Phil. 16(1888).55-57, suggests Domitian.

[36] Mart. 6.39.105 f.; cp. Friedländer-Magnus 1.244.

[37] Mart. 8.16.

[38] CIL. 6.1958; Blümner, *Tech.*, 1.39, fig. 13; 40, fig. 14; Platner 474.

[1] Cic. *Quinct*. 11.

[2] Juv. 1.99-101.

names to him when they called upon him or met him on the street. Petronius mentions a crier of lost children;[3] and there were also public officials called *praecones*, who attended certain magistrates and, as the vocal medium between them and the people, performed a variety of tasks dependent upon the office of him whom they served.[4] Some, for instance, summoned to court plaintiffs, defendants, and witnesses. It is doubtless this practice to which Martial refers when he says to Fabianus, whom he urges to keep away from the city: "Potes . . . nec pavidos tristi voce citare reos";[5] to judge from the context, he accuses court summoners of a tendency to become *delatores*, and enrolls them in his list of rogues.

With none of the foregoing, however, is the present account concerned; we are more interested in the auctioneer, the *praeco* to whom Juvenal referred the downtrodden provincial, that the latter might dispose of the few tattered effects that had not fallen into the hands of an extortionate governor.[6] A *praeco* of this sort was wont to set up a *hasta* in the forum[7] or public squares, at the cross-roads or street corners,[8] and sell to the gathered throng of common people[9] all kinds of cheap trumpery, *vilia scruta*;[10] old garments,[11] for instance, flagons, tripods, bookcases, caskets, and second-hand books by third-rate dramatists such as Paccius and Faustus.[12] Or if it were some ruined bankrupt whose goods were being put up under the sign of the spear,[7] the *praeco* would collect a less lowly throng perhaps, in regular auction-rooms[13] like the *atria Licinia* which Cicero mentions.[14] Public sales were advertised in advance, but all unfortunates were probably not so indifferent as the poor undertaker described by Petronius. He had once been able to dine like a prince, spilling more wine under his table than some people had in their cellars; but his business began to fail, and so fearing that

---

[3] Petron. 97 f.; cp. Plaut. *Merc.* 663 f.
[4] Mommsen, *Staatsrecht*, 1.347-350.
[5] Mart. 4.5.4; cp. Suet. *Tib.* 11.3.
[6] Juv. 8.95-97 and *schol.*
[7] Cic. *Off.* 2.83.
[8] *Id. Leg. Agr.* 1.7.
[9] Hor. *Epist.* 2.3.419.
[10] *Ib.* 1.7.65.
[11] Juv. 8.95.
[12] *Id.* 7.10-12.
[13] Cic. *Leg. Agr.* 1.7, *in atriis auctionariis*; cp. Juv. 7.7.
[14] *Cic. Quinct.* 12.25. Cp. Jordan 1.2.433; 1.3.331.21, 359.42 *vs.* Lanciani, *Ruins and Excavations*, 400 (Boston, 1897); Platner 460.

his creditors would surmise that he was going into bankruptcy, he nonchalantly advertised as follows, "Caius Julius Proculus will sell at auction some of his superfluous articles!"[15]

Cicero expresses his opinion of auctioneers in his oration for Publius Quinctius, when he chides Gaius Quinctius for associating with Sextus Naevius, "whom nature had endowed with nothing but a voice, to whom his father had bequeathed nothing but freedom." He was a *bonus vir*, the orator admits, a witty buffoon, and a civil auctioneer, but he lacked the training and culture, "ut iura societatis et officia certi patris familias nosse posset."[16]

The article *Praeco* in Smith's Dictionary of Greek and Roman Antiquities contains the statement that the contempt in which the office of *praeconium* was held is seen in Juv. 3.33, 7.6, and CIL. 1, 206.[17] An examination of these passages may prove interesting. The first in its context is as follows:

<pre>
29   Vivant Artorius istic (i. e. Romae)
          et Catulus, maneant qui nigrum in candida vertunt,
          quis facile est aedem conducere flumina portus,
          siccandam eluviem, portandum ad busta cadaver,
33   et praebere caput domina venale sub hasta.
</pre>

For verse 33 two interpretations are commonly suggested: either "to be sold up as a bankrupt," or "to sell slaves at auction." According to the former, the thought seems to be that "only dishonest men, who will stoop to the basest means, can thrive at Rome; those for instance, who after taking certain public contracts, embezzle the money received, put the greater part of their property beyond the reach of the law, and then go into bankruptcy to defraud the state." If this is the meaning, it would seem that the stigma implied in *facile est*, verse 31, holds over to verse 33, and does not necessarily attach itself to *conductores* in general, but merely to the individual scoundrels who entered into public contracts with the intention of cheating the state. Under the second explanation, contractors as a class are scorned and put on a par with dealers in slaves, who, as we have seen, were in very poor repute.[18] Friedländer and others, in their editions of Juvenal, adopt the rendering "sell slaves at auction," and observe that the line shows the calling of *praeco* to be despicable. Mayor's note on the verse itself suggests "is sold up"

---

[15] Petron. 38.
[16] Cic. *Quinct.* 11 f.
[17] W. Smith and G. E. Marindin in Smith 2.475 f.
[18] See pp. 29 f.

as a translation, but on Satire 7.6, he says: "How much the *praecones* were despised, appears from III 33 n. 157." There is great ambiguity here: his note on 3.33 is not concerned with *praecones*, and according to his interpretation of the line, any slur that is insinuated is certainly on the bankrupt, not on the auctioneer who sold his property *sub hasta*.

Although the embezzlement supposition is ingenious and fits the sense of verse 30, it is rather involved and requires much reading between the lines. The other has some support from the verses immediately following, which should undoubtedly be taken in close connection with 29-33:

> 34 Quondam hi cornicines et municipalis harenae
> perpetui comites notaeque per oppida buccae
> munera nunc edunt et, verso pollice vulgus
> cum iubet, occidunt populariter; inde reversi
> conducunt foricas.

There is the possible inference that men who had been trumpeters at gladiatorial contests in provincial towns and who had, threfore, doubtless been slaves themselves, might find incentives and opportunities to acquire wealth by becoming dealers in slaves, and so might sell them at auction, employ them as workmen for public contracts which required heavy or disagreeable labor,[19] or exhibit gladiatorial shows of their own. In this case, however, the subject of the infinitive would naturally be *mangones* or *mancipes*, rather than *praecones*.

But there is a third interpretation, an off-shoot of the first, which I have not noted in any of the editions, although it seems very obvious. According to this, the line refers to *delatores*, and the passage in full means: "men who can take up public contracts of the most degrading sort, inform against one another, and in this way furnish the emperor with those whose property may be sold at public auction to fill his coffers." This translates *praebere venale* literally and is substantiated by the scholiast, who explains *venale sub hasta* by the clause, "qui possunt a fisco vendi quasi debitores fisci."[20] The rendering is wholly in keeping with the immediate context: almost at this very point, Juvenal makes Umbricius exclaim, "What am I to do at Rome? I can not lie"; then after adding that he has no genius for flattering, fortune telling, or abetting murder, adultery, or theft, he ends this part of his discourse

[19] Cp. Trajan, Plin. *Epist.* 10.32 (41): "Ministeria quae non longe a poena sint . . . ad balineum, ad purgationes cloacarum, item munitiones viarum et vicorum."

[20] It is rather significant that the scoliast says *quasi*, and uses *fiscus* rather than *aerarium*, but cp. Tac. *Ann.* 6.2(8).1: "Et bona Seiani ablata aerario ut in fiscum cogerentur tamquam referret."

with the complaint, "Who finds favor now unless he is a confidante and has a mind boiling and seething with secrets which should never be revealed. ... Dear will he be to Verres who can accuse Verres at any time he wishes."[21] The thought reappears later in the same Satire, when as a climax to his invective against the Greeks, Juvenal through Umbricius accuses the race of being expert *delatores*.[22] Finally, special reference to the informer is included in parallel passages from both Juvenal and Martial.[23]

Since Juv. 3.33, therefore, is most probably an allusion to *delatores*; or may be primarily concerned with bankrupts or *mangones*; and if it refers to auctioneers at all, has to do specifically with only one class who dealt with slaves and gladiators,[24] it is scarcely fair to give it a place of prime importance in a generalization upon the estimate of the trade *praeconium*.

Now let us return to the second reference cited in the classical dictionary. This is Juv. 7.6; with the adjoining lines it reads:

```
5           Nec foedum alii nec turpe putarent
            praecones fieri, cum desertis Aganippes
            vallibus esuriens migraret in atria Clio;
            nam si Pieria quadrans tibi nullus in umbra
            ostendatur, ames nomen victumque Machaerae
10          et vendas potius commissa quod auctio vendit.
```

The satirist has been complaining of the neglect of literary men which has forced poets of renown to the point of keeping bathing establishments, running bakeries, and becoming auctioneers. While verse 5 does imply that there had been people who considered such callings beneath their notice, it also gives indication that in the first century after Christ, the attitude toward various occupations was changing. In view of the dishonesty that was rife among favored freedmen and even in higher circles, men of worth were doubtless beginning to learn that humble employments could offer at least an honest livelihood. As Juvenal says: "If there should be no sign of a single cent for you in the shady grotto of the Muses, you would adore the name and calling of Machaera,[25] and prefer to open an auction and sell what it has to offer to a crowd of bystanders. ... This is better than to declare before

[21] Juv. 3.41-54.
[22] *Ib.* 113-125.
[23] Mart. 4.5, cp. 3.38; Juv. 7.1-16.
[24] Cp. Mart. 6.66; Juv. 3.157 f.
[25] Most editors agree that Machaera was a *praeco* of Juvenal's time, but Weber (quoted in Mayor's edition) compares it to Gr. μάχαιρα and thinks it may mean "cook."

a judge, 'I have seen' what you have not seen, as knights who hail from Asia do."[26] Although he adds later that thanks to the patronage extended by Hadrian, the hopes of literary men have revived because none henceforth will be forced to endure *studiis indignum laborem*,[27] his words should scarcely be considered a special disparagement of the tasks he has just mentioned; for *indignum laborem* from his point of view would no doubt refer to any exertion whatever that would be likely to divert poets from their pursuit of the Camenae. Taking the extract as a whole, therefore, it appears to be rather to the advantage of the *praeco* than otherwise.

In commenting upon Juv. 7.6, a line which beyond a doubt alludes to auctioneers, Mayor notes that *praecones* were not eligible to the rank of decurion, so long as they followed their profession. This information he obtained from a portion of the Lex Julia Municipalis which happens to be the third of the passages mentioned at the beginning of this discussion. The Latin is: "Neve quis, quei praeconium dissignationem libitinamve faciet, dum eorum quid faciet, in muni / cipio colonia praefectura II vir(atum) IIII vir(atum) aliumve quem mag(istratum) petito neve capito neve gerito neve habeto, / neve ibei senator neve decurio neve conscriptus esto neve sententiam dicito."[28] Now *praeconium* here may of course have its general signification and mean all persons who held the office of *praeco* of any kind. If so, then *dissignationem* too should be generic, including all who served as *designator*, such as the master of ceremonies at funerals,[29] the usher at the theatre,[30] and the umpire at public spectacles.[31] However, its close connection by *-ve* with *libitinarii* leaves no doubt that its use in this case is specific, and that it denotes the *designator* in his relation to funerals. It must be permissible, therefore, to take *praeconium* also in a restrictive sense; and so Tyrrell, as was evidently Mayor's intention too, refers it to the auctioneer, and explains that he was apparently regarded with detestation like modern pawnbrokers and usurers, as trading upon the misfortunes of others.[32] But when the word admits a choice of meanings,

---

[26] Juv. 7.8-11, 13 f. Cp. Petron. 116, see p. 39, n. 65.
[27] Juv. 7.17.
[28] CIL. 1.206. 94 ff.
[29] Hor. *Epist.* 1.7.6 and *schol.*; Sen. *Benef.* 6.38.4; Tertul. *Spect.* 10.
[30] Plaut. *Poen., prol.* 19 f.
[31] Cp. Dig. 3.2.4.1.
[32] Tyrrell and Purser, *Correspondence of Cicero*, 4.419 (London, 1894), note on *Epist.* 6.18.1. Cp. Post on Mart. 1.85.

it seems rather incongruous to assume at once that it pertains to the *praeco* of the auction-room, especially when the only other classes mentioned in the same clause are funeral marshals and undertakers. The belief that consistency would be maintained in a formal law suggests that there were *praecones* whose duties were obituary. The reason for their being debarred from participation in political life would then be the same as for *designatores* and *libitinarii*. Furthermore there are the clear statements of Festus[33] and Varro[34] that the services of a *praeco* were employed to summon the participants in a public funeral. In the words of Marquardt, who sums up and expands their evidence: "Die Aufforderung zur Theilnahme an jedem solennen Leichenzuge erging durch einen öffentlichen Aufruf (davon *indictivum funus*), bei welchem der Herold mit den Worten einlud: 'Ollus Quiris leto datus. Exsequias, quibus est commodum, ire iam tempus est. Ollus ex aedibus effertur.'"[35] There can be little doubt that it was these *praecones* attendant upon funerals, and not ordinary auctioneers, whom the Lex Julia Municipalis declares ineligible to become decurions in municipalities, colonies, and prefectures. The ban was evidently put upon them only because of their connection with the dead. Should they resign their office, it appears that no stigma attached to them, but that they could be elected to the highest magistracies; for Cicero, after inquiring into the law, wrote to a friend: "Rescripsit eos, qui facerent praeconium, vetari esse in decurionibus; qui fecissent, non vetari."[36] While they followed their profession, those at Rome were doubtless under the direction of the funeral contractors, with headquarters at the Temple of Libitina.[37] That the duties of associated officials might be combined, at least in small towns, is testified by a sepulchral inscription, which also adds strong evidence for consigning the *praeco* now under dispute to funereal employment. It reads:

C.MATIENI.C.F.OVF
OVICVLAE
ANNORUM.xxvii
PRAECO.IDEM.DISSIGNATOR.[38]

[33] Fest. 106 M, 254 M.
[34] Varro *Ling.* 5.160; 7.42.
[35] Marquardt 1.351.
[36] Cic. *Epist.* 6.18.1.
[37] Cp. Marquardt 1.384 f.
[38] CIL. 10.5429.

Although it seems necessary, then, to discard several of the passages that are commonly accepted as alluding to auctioneers or including them, there are still a number from which something may be gleaned about the character of this class of people and their condition in life. Loquacity was a noticeable trait. Cicero terms an auctioneer pert, *dicax*, and adds that one of free birth would take advantage of his *libertas* to indulge in special freedom of speech.[39] Horace intimates that the chatter of *praecones* was enticing and persuasive, having the power to attract a crowd with the lure of wonderful bargains to be obtained.[40] In Juvenal's estimation, this talkativeness surpassed that of the grammarian, rhetorician, and advocate, but was doomed to fall before a woman's art.[41] According to Martial, it was likely to develop into garrulousness and end in extreme stupidity. He instances a facetious *praeco* who was attempting to dispose of some highly cultivated fields and magnificent acres of land on the outskirts of the city: " 'Whoever thinks that Marius is forced to sell' comments the auctioneer, 'is very much mistaken; he is not in debt, but quite the contrary, and has money out at interest.' 'What's the matter then?' someone asks. 'Why he has lost everything here,' is the ready answer, 'all his slaves, flocks, and crops, consequently he doesn't like the place.' "—Nor does anyone else, as Martial remarks in conclusion, and Marius's ill-fated farm still clings to him.[42] But some *praecones* could be very effective talkers, the poet admits in another epigram: "Two praetors, four tribunes, seven advocates, ten poets," he says, "were recently asking a certain old gentleman for the hand of his daughter in marriage. Without a minute's hesitation he gave her to Smooth-talk, the auctioneer!" The epigrammatist does not take the responsibility of condemning him, but closes with the question: "Tell me, Severus, was he altogether a fool?"[43] There was of course a motive behind the father's choice. This may easily be traced to the suitor's financial standing; for even as early as the time of C. Laelius Sapiens, a certain auctioneer, Gallonius by name, had been serving novel dainties upon his table and living in the lap of luxury.[44] His name had become a byword and is recalled by Cicero and

---

[39] Cic. *Quinct.* 11.
[40] Hor. *Epist.* 2.3.419 f.
[41] Juv. 6.438-440; cp. Fulgent. *Myth.* 1(p. 23 Muncker).
[42] Mart. 1.85; cp. 6.66.
[43] *Id.* 6.8.
[44] Cic. *Fin.* 2.24.

Horace.[45] Under Domitian even such stupid specimens as the above mentioned Marius had employed were making an enviable livelihood, so that Martial, in an extremely modern tone, makes a strong, if sarcastic, plea for vocational training to settle that momentous question, "After College What?—For Boys." Don't send your son to the grammarians and rhetoricians, he admonishes Lupus, to have him waste his time over Cicero or Virgil. If he shows an inclination to write verses, disinherit him. If he wishes to learn lucrative arts, have him taught music that he may perform on the lyre or the pipes; but if he is dull, make him an architect or an auctioneer![46] One of Trimalchio's friends may have been influenced by similar advice; for although he hoped to make his boy something of a jurist, he decided, if the child recoiled from this, to have him learn the trade of barber, auctioneer, or advocate at least—something which he could carry to the grave with him.[47]

According to Friedländer on Mart. 1.85.1, "Das Gewerbe des Ausrufers bei Auktionem stand dem des Spassmachers nahe und darum in Missachtung."[48] If we were to judge entirely from the *praecones* who come under the sting of Martial's ridicule, we might be inclined to believe that this characterization is very near to the truth; but Horace presents quite a different type in the Volteius Mena of one of his Epistles.[49] He represents him on his own declaration as an auctioneer, a man of modest circumstances and of blameless reputation, who enjoyed a home of his own and agreeable friends of humble rank like himself. He liked to resort to the games on holidays and to the sports of the Campus Martius after business was over, but he knew how to work and play, make money and spend it, each at the proper time. Unfortunately this *praeco* resigned his independence and became the subservient client of the famous orator L. Marcius Philippus,[50] who while walking across the Forum to his home on the Carinae early one afternoon, saw Volteius leisurely sitting in a barber shop, already shaved and quietly cutting his own nails; Philippus was at once attracted, apparently by his appearance of ease and contentment. We recall how Mena, although resisting

---

[45] *Id. Quinct.* 94; Hor. *Sat.* 2.2.46-48.

[46] Mart. 5.56. Cp. 9.73 and see p. 58. For further allusions to the lack of patronage granted to letters, eloquence, and learning in general, cp. Petron. 83, 88, 116; Juv. 7. 1 ff.

[47] Petron. 46.

[48] Cp. Cic. *Quinct.* 11.

[49] Hor. *Epist.* 1.7.46 ff.

[50] Cp. Cic. *Brut.* 173; *De Orat.* 3.4.

the advocate's advances at first, finally yielded to the glamor of a client's life, but later begged to be returned to his former condition, after he had made a dismal failure of farming. This anecdote from Horace's pen praises contentment with one's lot[51] and shows that that of the *praeco* could be far from undesirable. It gives a pleasing and no doubt accurate picture of what must have been the life of the ordinary auctioneers who attended to business and did not bother about the shallow proprieties of high society, but kept themselves morally upright, and found joy and satisfaction in living. They may have been exposed to snobbishness or to the temptation of parasitism, but they may also have resisted them both and in leading their own lives well, have met with the respect of their fellow-men, being measured by their character, not their profession. Hence it was, doubtless, that Horace's father, who was in all probability connected with auctions himself, thought it worth while to give his son an education of culture and refinement, even though it should fall to his lot to become only an auctioneer or collector. In the words of the poet:

> Nec timuit sibi ne vitio quis verteret olim
> si praeco parvas aut, ut fuit ipse, coactor
> mercedes sequerer: neque ego essem questus.[52]

Then follow those loyal words that were no doubt echoed by many another worthy son in similar station:

> Nil me paeniteat sanum patris huius.

XXII

SUTORES   CERDONES

It would seem only natural to find that in primitive Rome tanners had for some time engaged in all varieties of the leather business, including shoemaking; yet according to Plutarch, *sutores* were incorporated from the first in a separate *collegium*.[1] Literary allusions to this class are comparatively numerous. Possibly because they plied a very familiar trade, they are not uncommonly mentioned generically for working men or for the common people in general. Cicero, for instance, designates a popular assembly at Pergamum as a gathering of *sutores et zonarii*—and a virtueless set he considers them; incapable of upright judgment, and abjectly subservient to the demagogue who could best

---

[51] Cp. Hor. *Sat.* 2.6.79-117, a fable of the same purport on "The Town and Country Mouse."

[52] Hor. *Sat.* 1.6.85-87; cp. Suet. *Vita Hor.* 44 (Reifferscheid).

[1] See p. 1.

sate their appetites.² Horace twice refers to shoemakers as typical *opifices*; he adds no disparaging remarks upon their character, but rather calls attention to their skill.³ Juvenal uses the same figure. It occurs in a passage of stinging derision in which a drunken brawler, ready to offer insult, to pick a quarrel, exclaims:

> Quis tecum sectile porrum
> sutor et elixi vervecis labra comedit?
> nil mihi respondes? aut dic aut accipe calcem.
> ede ubi consistas, in qua te quaero proseucha?⁴

The sentiment expressed may not have been Juvenal's own; and when we consider the source from which it sprang, we must conclude in all justice that it may not necessarily have been the view of fair-minded men in his time, but was perhaps a reflection of the common republican attitude surviving in a bully and worthless reprobate, who probably had nothing else to his credit but birth. It is to be noted that the intended calumny in classing the wayfarer with *sutores* did not prove strong enough to produce the desired effect, and resort had to be made to more obvious slander.

Friedländer comments that *sutor* in the verses just quoted is used contemptuously as in Juv. 4.153, 8.182; Mart. 3.16.1, 59.1, and 99.⁵ Now it happens that neither Juv. 4. 153:

153   Sed periit postquam cerdonibus esse timendus
            coeperat; hoc nocuit Lamiarum caede madenti,

nor 8.182:

179   Quid facias talem sortitus, Pontice, servum?
            nempe in Lucanos aut Tusca ergastula mittas.
            at vos, Troiugenae, vobis ignoscitis, et quae
182   turpia cerdoni, Volesos Brutumque decebunt,

Cerdones

contains any specific reference to a shoemaker! *Cerdo*, which is sometimes interpreted "cobbler"⁶ on the strength of the Martial passages cited, should in all probability be written with a capital;⁷ for it was a Greek proper name commonly

---

² Cic. *Flacc.* 17.
³ Hor. *Sat.* 1.3.124-133; 2.3.106.
⁴ Juv. 3.293-296.
⁵ Friedländer's notes on these passages are all-inclusive and inconsistent.
⁶ Harper's Lex. and many commentators.
⁷ So Conington opines; cp. his note and text for Pers. 4.51, "Tollat sua munera Cerdo."

applied, like Dama,[8] to slaves and freedmen;[9] the word also appears to have been employed at times to represent the slave class or rabble.[10]

The Digest, for instance, uses it as a slave name typifying a class; observe the formula, "Cerdonem servum meum manumitti volo, ita ut operas heredi promittat."[11] In the light of history, the only plausible interpretation for *cerdonibus* of Juv. 4.153, which concerns the death of Domitian, is that it designates the ranks of slaves or of those who had been born in slavery; for Suetonius enumerates as the Emperor's assassins: "Stephanus, Domitillae procurator . . . Clodianus cornicularius et Maximus Parthenii libertus et Satur decurio cubiculariorum et quidam e gladiatorio ludo."[12] The poet's words therefore evidently mean: "Domitian met his fate when he began to be an object of alarm to Cerdos (*i.e.*, slaves and freedmen). This was the undoing of one who was reeking with the blood of Lamias (*i.e.*, noblemen)."[13] Only by spelling *Cerdonibus* with a capital is a proper balance obtained with *Lamiarum*. In each case, the poet chooses a name which stands for a class; it is quite natural that the allusion should be more specific in the second part: the name of some prominent noble who had suffered at Domitian's hands would doubtless be known to everyone, but it is not so likely that the identity of the slaves and freedmen who had been accomplices in the Emperor's murder would be a matter of common knowledge. The same contrast is drawn in Juv. 8.182; for *cerdoni* there is plainly just a synonymous repetition of *servum* in verse 179. The import of the quotation, then, is: "You haughty descendents of the Trojans make excuses for yourselves, and what is a disgrace for slave Cerdo will be deemed becoming for Lords Volesus and Brutus."

Κέρδων in Greek is expressive of knavish cunning;[14] compare κερδώ, "the wily one," "thief"; κερδοσύνη, "cunning," "craft," "shrewdness"; κέρδεα, pl., "cunning arts," "wiles," "tricks." The word in Latin, says Duff in his comment on Juvenal 8.182, "is clearly used as a

[8] Hor. *Sat.* 1.1.101 and *schol.*; 1.6.38; 2.5.18, 101; 2.7.54; Pers. 5.76, 79; Petron. 41; Mart. 6.39.11; 12.17.10.

[9] Demos. *Nicostr.* 1252; Petron. 60; CIL. 2.4970.130; 4. 6867, 6868; 5.5300; Dig. 38.1.42.

[10] Cp. *schol.* on Pers. 4.51; *id.* Juv. 4.153, 8.182. We may compare our common "Tom, Dick, and Harry," or, as Conington (see n. 7) suggests, the "Hob and Dick" of Shakespeare's *Coriolanus*.

[11] Dig. 38.1.42. Compare "John Doe" in modern legal usuage.

[12] Suet. *Dom.* 17.1 f.

[13] Cp. Juv. 6.385; Suet. *Dom.* 1.3; 10.2.

[14] Cp. *schol.* on Pers. 4.51; *id.* Juv. 4.153, 8.182; Fest. 56 M.

contemptuous sobriquet for the class engaged in small trade and handicraft."[15] Even if "clearly" should be stricken out, we could not accept the statement without restriction. There is little likelihood that the term was applicable indiscriminately to the "class engaged in small trade and handicraft," the definition in the Thesaurus, *operarius, opifex infimi generis*, is probably more accurate.[16] Judging from the etymology of the word and from its use in the references which we have examined, it appears to have been employed primarily as a proper name for a slave or freedman who was clever in turning his efforts along any line into honest (?) or, more frequently perhaps, dishonest gain. Especially in point is Petronius's allusion to three slaves of Trimalchio's, one of whom, their master said, "was called Cerdo; another, Felicio; the third, Lucrio."[17] Because of the common use of Cerdo as a slave name, it was sometimes made to typify, as we have seen, the servile class in general, especially the unscrupulous rabble. Finally, just as Tempe, for instance, could be applied to any beautiful vale,[18] so Cerdo may have been used occasionally to designate a slave workman, hence the definition in the Thesaurus, which has been quoted above. There is a fragment of a pillar at Pompeii which might throw light on this point if the inscriptions on its four sides were more intelligible. The following words, of interest here, can be discerned: CIL. 4.6867, CERDO SODALIBUS; 6868, CERDO HIC DIDIT; 6869, CERDO CERDONIBVS / SAL.; 6871, CERDO HI(C); 6877, OPERARI(I)S PANE(M) / DENARIV(M). Since this column was excavated at Boscoreale, it may bear witness to a *sodalitas* which had been organized by the slave workmen of the Villa. According to all indications, therefore, any implication involved in *Cerdo* is concerned with its relation to slaves (or freedmen); certainly there is no definite proof among the references which we have noted that it signifies "cobbler" or "shoemaker" specifically. In Spon's *Misc.*, page 221, mentioned by Jahn and Conington on Pers. 4. 51, it is coupled with *faber*; and in the inscriptions just quoted, it has some connection with

[15] So also Post on Mart. 3.99.

[16] Cp. *artifex sordidus*: Weise, *Griechischen Wörter im Latein*, 375 (Leipzig, 1882); Saalfeld, *Tensaurus Italograecus*, 262 (Wien, 1884). But Weise 202 also refers the word specifically to shoemakers.

[17] Heseltine, in his 1913 translation of Petron. for the Loeb Series, can surely not be right in assigning these names to the images of the *lares*, which were apparently *two* in number, as usual, and were brought in by *two* of the slaves. The "veritable image" of the freedman Trimalchio was doubtless the third slave, "Gain," or "Luck," or "Profit," whichever it was that carried the wine around.

[18] Cp. Virg. *Georg.* 2.469; Ov. *Am.* 1.15; *Fast.* 4.477.

bread and *operarii*; we may recall also the use of *operas* in the lines previously cited from the Digest. At times no doubt the main emphasis involved, not "craftsmen," but "crafty men."

Martial presumably had this point in mind in the three passages which seem to have given rise to the interpretation "cobbler," "shoemaker"; namely,

3.16.	1	Das gladiatores, sutorum regule, cerdo,
		. . . .
	6	nunc in pellicula, cerdo, tenere tua,
3.59.	1	Sutor cerdo dedit tibi, culta Bononia, munus, fullo dedit Mutinae: nunc ubi copo dabit?
3.99.	1	Irasci nostro non debes, cerdo, libello. ars tua non vita est carmine laesa meo. innocuos permitte sales. Cur ludere nobis non liceat, licuit si iugulare tibi?

There is every indication that the term under discussion should be capitalized in all three epigrams.[19] Since *sutor*, according to the lexica, signifies both "shoemaker" and "cobbler," the combination *sutor* (*sutorum*) *cerdo* in 16 and 59 is redundant; and the context of 99, especially verse 2 and *iugulare* in verse 4, proves almost conclusively that it refers to the same person as the others. Proper names occur much more frequently than common nouns as vocatives in the opening and closing verses of Martial's epigrams, and the wording of 16.1 particularly demands a *nomen*. It is not surprising that Martial does not name the *fullo* and the *copo* in 59; by specifying the *sutor*, he makes him of prime importance, and shows a connection between 59 and 16. Furthermore, the name was probably chosen to imply that this particular *nouveau riche* had been not merely a plebeian *sutor*, but originally a slave, and one whose wealth perchance had been accumulated by questionable means. This use of an appellative with double and appropriate meaning is altogether characteristic of Martial.

The epigrammatist would have found the proprieties better observed, it seems, if shoemakers had obeyed that proverbial admonition attributed to Apelles: "Ne supra crepidam sutor iudicaret."[20] "You are intoxicated," he exclaimed to a Cerdo who had exhibited a gladiatorial combat; "for never in your sober senses would you come to the point

---
[19] Duff in his note on Juv. 8.182 inclines to this opinion.
[20] Plin *Nat.* 35.84 f.

of wishing to take your amusement at the expense of your hide! You have had your sport to your cost, but take my advice, and remember henceforth to 'stick to your last.' "[21] It was apparently to excuse himself for this piece of pleasantry[22] that he wrote 3.99, declaring apologetically: "Ars tua non vita est carmine laesa meo"; that is (since *ars tua* and *iugulare* (v. 4) are presumably synonymous[23]), "it is your avocation, not your vocation, that I have satirized in my poem."

Yet shoemakers evidently did not "stick to their last." They appeared upon the street to hail returning travelers with a kiss of greeting, regardless of the fact that their lips had just been in contact with leather.[24] From their ranks, apparently, came one who, entering the literary arena, dared to criticize Martial's verses and write others of his own[25]—small wonder is it that *sutores* met with little favor from the epigrammatist's pen! Another of their number, or it may have been the same, became an envied *dominus* living in luxury on a splendid country estate at Praeneste. A slave's quarters there would have been too good for him, declares Martial, yet he had actually inherited the place from his patron. "Ah," the poet concludes with a truly modern flavor, "what fools my parents were to give me a liberal education! What good are grammarians and rhetoricians to me? Break my trifling pens and tear up my poems, Thalia, if a boot can give all that to a cobbler."[26]

"All that?"—Yes, and more; for under Nero, Vatinius, a *sutor* of Beneventum, rose to a position of influence, wealth, and power.[27] It

[21] Mart. 3.16.3-6.

[22] And *Id*. 3.59.

[23] I.e., *ars* = *ars gladiatoria* but does not, I think, include *ars sutoria* as Post holds. Strictly speaking, of course, the man who merely provides money for a show would not be said to follw the *ars gladiatoria*; but if 3.99 does refer to 3.16 and 3.59, as is highly probable, the thing that is criticized in these two epigrams is the giving of gladiatorial exhibitions; and in both 3.16.2, 4 f. and 3.99.4, Martial's diction playfully confuses the donor of the shows with the gladiator himself.

[24] Mart. 12.59.6 f.; cp. 3.16.6; 6.64.31; 9.73.1. See pp. 22, 75 f.

[25] Mart. 6.64.

[26] *Id*. 9. 73. Martial 5.56 expresses the same sentiment in connection with the musician, the architect, and the auctioneer; cp. Petron. 46. See p. 52.

[27] Cp. Porph. on Hor. *Sat*. 1.3.130. He records that Alfenus Varus, a *sutor* of Cremona, closed his shop and went to Rome; there he became intimate with Sulpicius the jurisconsult, and rose to such high position, that he won the consulship, and at his death was honored by a public funeral. Acron says that the consul was a "son of a shoemaker." But see p. 91, n. 36. According to one report chronicled by Suet. *Vitel*. 2.1, this Emperor was the great-great-grandson of a *sutor veteramentarius*.

was not, however, the trade which he had learned in a *sutrina taberna* that had set him on his towering eminence and roused the people's disdain and hatred for him, but his physical deformity and moral depravity; to wit, his unusually long nose and his sycophantic habits.[28] The former gained him admission to the imperial court as a buffoon,[29] the latter formed the ladder by which he climbed. In mockery of him apparently, cheap goblets with very long spouts were known as *calices Vatinii*. Martial tags one of them with the following expressive couplet:

> Vilia sutoris calicem monimenta Vatini
> accipe; sed nasus longior ille fuit.[30]

No doubt the epigrammatist's scorn for Vatinius was shared by many, but it is to be questioned whether his animosity against shoemakers in general was shared by all. We may readily infer that much of his satire was aroused by individual cases on personal grounds. Moreover, the *sutores* of his poems are slaves[31] or freedmen, and rogues besides; but his lines do not convey the idea that they were shoemakers and *therefore* rogues, but rogues who *happened* to be shoemakers.

There were respectable *sutores*, presumably, who played no insignificant part in town life. At Athens, according to Lysias, the cobbler's, the perfumer's, and the barber's offered favorite rendezvous;[32] since the same was true at Rome of the last two,[33] it was probably so with the first. The headquarters of the shoetrade is referred by Platner to the Argiletum.[34] His belief seems to be based upon a single passage from Martial:[35]

> Tonstrix Suburae faucibus sedet primis,
> cruenta pendent qua flagella tortorum
> Argique Letum multus obsidet sutor.

This is rather doubtful authority for such a sweeping assertion. The reference at most can be concerned only with the upper end of the Argiletum where it meets the Subura. But it is more probable that the

---

[28] Tac. *Ann.* 15.34; *Hist.* 1.37; *Dial.* 11; Dio 63.15.
[29] Cp. Suet. *Tib.* 61.6.
[30] Mart. 14.96; cp. 10.3.4; Juv. 5.46-48 and *schol.* Furneaux, in his comment on Tac. *Ann.* 15.34, suggests that Vatinius may have made the cups that bore his name; but this does not seem a natural inference in view of Martial's distich.
[31] *Sutores* were sometimes found among domestic slaves, cp. Petron. 68.
[32] Lysias *Orat.* 24.20; cp. Demos. *vs. Phorm.* 13.
[33] See p. 73.
[34] Platner 459, cp. 457. Jordan's statement 1.3.328 is not so strong, but he, too, quotes only the one reference; cp. 1.2.452, where he consigns the headquarters to the Subura and the Vicus Sandaliarius.
[35] Mart. 2.17.1-3.

qua clause is a periphrasis for *qua incipit Vicus Sandaliarius*; for this street, if the location assigned to it by topographers be the correct one,[36] branched off from the Argiletum near its juncture with the Subura; and since it practically took the place of a continuation of the main thoroughfare, if we consider the narrow entrance between it and the Subura proper as the *Suburae fauces*, it may well be said to have "blocked the Argiletum." There is apparently no objection therefore to applying Martial's words to the Vicus whose name naturally designates it as the center of the shoe business.[37] The street seems to have been in good repute: Augustus erected a statue of Apollo there,[38] and the *magistri vici* are shown by inscriptions to have been especially active and generous.[39] Besides a Sandalmakers' Street of some prominence, there was a Shoemakers' Hall which is mentioned in literature and inscriptions in connection with a certain religious celebration, the *tubilustrium*.[40] Here on March and May twenty-third the *sacrorum tubae*, which according to Fowler were to be used for assembling the *comitia curiata* on the next day,[41] were purified by the sacrifice of a lamb.[42] The festival was the occasion for a half holiday.[43] Why the Atrium Sutorium was chosen for the rite is open to conjecture; possibly it was merely on account of its size or convenient location,[44] although there is the natural inference that *sutores* may have been concerned with the celebration in some special capacity. At all events our evidence proves that shoemakers exerted an influence in their community even beyond their own peculiar province.

XXIII

TABERNARII

With commerce and trade making great strides under the Empire, as wealth increased and extravagance grew rife, Rome became a city of

[35] Jordan 1.3.329; Platner 448.

[37] See n. 34.

[38] Suet. *Aug.* 57.1. This was doubtless because of his interest in booksellers, who also had shops there during the Empire. See pp. 62 ff.

[39] CIL. 6.448, 761.

[40] Varro *Ling.* 6.14; Fest. 352 f. M; *Fast. Praenes.*, CIL. 1, p. 315.

[41] W. W. Fowler, *Roman Festivals*, 64 (New York, 1899).

[42] Fest. *l. c.*

[43] CIL. 1, p. 315.

[44] The location of the Atrium is not known. Jordan 1.2.452 and Platner 459 think that it may have been on the site of the Forum Transitorium; Mommsen, *Arch. Zeitung* 5(1847).109, and Gilbert 1.144 argue for the Palatine; Mommsen, CIL. p. 369, identifies it with the Atrium of Minerva.

shops of every description. Juvenal paints their vicinity as unattractive by night; for their shuttered fronts were chained and barred, so that the streets were dark and were frequented by thieves and ruffians.[1] But by day *tabernae* were scenes of color and animation; they were open to the air in front and sometimes on the side; their pillars might be hung with flagons, books, or other wares, or covered with advertisements of the commodities on sale within.[2] True to the innate Italian tendency to conduct any operation of daily life under the open sky, the inconsiderate shopkeeper had overrun the whole city, Martial declares, leaving no trace of a threshold where a threshold naturally should be.[3] Domitian finally passed a law forbidding encroachment on the public thoroughfare.[4] The poet's epigram on this subject is particularly vivid and illuminating:

> Abstulerat totam temerarius institor[5] urbem
>   inque suo nullum limine limen erat.
> iussisti tenuis, Germanice, crescere vicos,
>   et modo quae fuerat semita, facta via est.
> nulla catenatis pila est praecincta lagonis
>   nec praetor medio cogitur ire luto,
> stringitur in densa nec caeca novacula turba
>   occupat aut totas nigra popina vias.
> tonsor, copo, cocus, lanius sua limina servant.
>   nunc Roma est, nuper magna taberna fuit.[6]

Of the business sections, the Argiletum is popularly claimed to have been the center of the book and shoe trade.[7] We have already questioned whether the evidence is strong enough to uphold the last part of this tenet,[8] and Professor Tracy Peck also takes exception on the same grounds to a sweeping assertion about the book-trade. It rests, he says, on three references from Martial, two of which, at least, may refer to the same place.[9] After reviewing the available sources, in which he finds

---

[1] Juv. 3.302-304.
[2] Hor. *Sat.* 1.4.71 f.; *Epist.* 2.3.372 f.; Mart. 1.117.11 f.; 7.61.5; Juv. 8.168, see p. 15, n. 12.
[3] This may refer also to the practice of building wooden booths out over the sidewalk, cp. Typaldo-Bassia 24, Friedländer-Magnus 1.5 f.
[4] Friedländer, in his note on Mart. 7.61, assigns the law to the autumn or winter of 92.
[5] For a discussion of this word see p. 24.
[6] Mart. 7.61. See pp. 11, 15, 24, 28, 89.
[7] Becker 335; Jordan 1.3.328.15; Post on Mart. 1.2.8; Platner 173, 457.
[8] See pp. 59 f.
[9] Cp. Jordan, *Hermes* 4(1870).232 f. See p. 63.

mention of the sale of books in the Sigillaria,[10] the Vicus Tuscus,[11] the Forum,[11] and especially the Vicus Sandaliarius,[12] where according to Galen most bookstores were in his day, he concludes that the trade was widely distributed and that at least in the second century of our era, it was particularly prominent, not in the Argiletum, but in the Vicus Sandaliarius.[13] Professor Peck's main premise and ultimate conclusion are undeniably true, but we find, after a close scrutiny of the pertinent passages, that several slightly different or additional inferences may be drawn.

Martial has seven epigrams relating to booksellers. Four of them are of no topographical interest; three of the four disclose the names of Quintus Valerianus Pollio[14] and Tryphon.[15] The latter was apparently the editor of Quintilian also; to his accurate work the grammarian bears testimony in the dedicatory preface of the *De Institutione Oratoria*. Turning to the three excerpts from Martial, which have acquired perhaps undue importance, we read:

1.2.7	Libertum docti Lucensis quaere Secundum limina post Pacis Palladiumque forum,
1.3.1	Argiletanas mavis habitare tabernas, cum tibi, parve liber, scrinia nostra vacent,
1.117.9	Argi nempe soles subire Letum: contra Caesaris est forum taberna . . . .
13	illinc me pete. Nec roges Atrectum— hoc nomen dominus gerit tabernae—: de primo dabit alterove nido. . . . . (Martialem).

The second clearly states that Martial's books could be purchased in more than one store of the Argiletum. The third also plainly refers to the same street and locates the shop of Atrectus opposite the Forum of Caesar. There is doubt as to which forum is meant, but it seems most natural to understand *Caesaris* in the usual manner as relating to the reigning Emperor[16] and designating therefore the Forum Palladium or

[10] Gell. 2.3.5; 5.4.1.
[11] Hor. *Epist.* 1.20.1 and *schol.*
[12] Gell. 18.4.1; Galen 19.8 (Kühn).
[13] Peck, Class. Phil. 9(1914).77 f.
[14] Mart. 1.113.5 f.
[15] *Id.* 4.72.2; 13.3.4.
[16] Most editors incline to this view. Cp. also Jordan, *Hermes* 4(1870).232; Hülsen, *Rheinisches Museum* 49(1894).630.

Transitorium, which, though begun by Domitian, was finished by Nerva and was afterward associated with his name.[17] Friedländer, however, favors the Forum Julium, which was also called *Caesaris*.[18] It is unnecessary to attempt to settle the dispute; for the same general locality at the beginning of the Argiletum may be involved in either case.[19]

The first extract quoted above does not expressly mention the district in question. Furthermore, there are grounds for disputing the theory that Secundus (1.2.7) and Atrectus (1.117.13) were one and the same: in the first place, the context implies that the former sold choice and rare editions of parchment, while the latter dealt in the more common papyrus rolls;[20] then too, it has been argued, as we have noted, that *Palladium* and *Caesaris* may not signify the same forum; finally, the prepositions may indicate that a physical inability would be involved, unless Atrectus Secundus had two establishments and divided his energies between them as Martial distributed his name. If the point of view is from the Forum Romanum Magnum, as reason would lead us to infer in 1.2, and the context of 1.117 (especially verse 6, "Longum est, si velit ad Pirum venire," and the use of *subire*, verse 9), then *post* (= "behind") and *contra* (= "opposite," *i.e.*, "in front of") would designate opposing directions.[21] Moreover, it would be illogical to suppose that the location would be changed, if one should wish to adopt Stephenson's comment and consider *post* from the position of Martial's house, which was probably on the west slope of the Quirinal.[22] Surely the front and rear of the imperial fora would be determined by their relation to the great Roman Forum; the phrase "behind the Temple of Peace and the Forum Palladium" would therefore describe a quarter that would remain immobile, regardless of the chance position of a directing agent. Now *limina post Pacis Palladiumque forum* seems a

---

[17] Jordan 1.2.449-453; Gilbert 3.232 f.; Platner 282-284.

[18] Friedländer on Mart. 1.117.10. Cp. Jordan 1.2.437-441; Gilbert 3.225-227; Platner 275 f.

[19] See p. 64.

[20] I find that Friedländer, on Mart. 1.2.7, resorts to this argument.

[21] If it could be *proved* that *contra Caesaris . . . forum* refers to the upper end of the Argiletum, then the sites indicated by *contra* and *post* might coincide; but the points adduced above would have to be considered.

[22] Cp. Mart. 1.108.3; 1.117.6.

needlessly full and even inaccurate expression[23] to designate only the Argiletum,[24] but may with good reason be regarded as evidence that in Martial's day, too, a favorite stand for bookdealers was the Vicus Sandaliarius, which, according to topographers, branched off from the Argiletum *behind* the Forum Palladium, and ran directly *behind* the Forum of Peace in a course almost parallel to it.[25]

We have an intimation that this same Vicus was a book district even as early as the time of Augustus; for when the Emperor was setting up images of gods in several streets, he chose for the Vicus Sandaliarius a statue of Apollo,[26] whom, in his desire to turn the minds of men to the pursuits of peace and literature, he was raising to the first place of prominence in Roman religion.[27] Further information on bookshops of the Augustan period is revealed by Horace's words to his book:

> Vertumnum Ianumque, liber, spectare videris,
> scilicet ut prostes Sosiorum pumice mundus.[28]

The lines apparently indicate two sections just off the Forum, one in the vicinity of the statue of Vertumnus in the Vicus Tuscus;[29] the other near the shrine of Janus Geminus, which references in literature consign to the lower end of the Argiletum,[30] a district corresponding perhaps to that just noted in Martial 1.117.9 ff.[31] The Sosii, says Porphyrio, were two brothers who were well-known bookdealers of the time.[32] From the verses quoted above, it would appear that their firm did business on both sides of the Forum.[33]

[23] It might designate a district about 185 m. in breadth, since the dimensions given for the Forum of Nerva are 40 x 120 m., and for the Forum of Vespasian 145 x 85 m. (Platner 281 f.).

[24] Of course, as the Argiletum ran diagonally, it might roughly be said to have been "behind" both fora, but there is no need to suppose that *post* is used so loosely, when a close and natural rendering of the Latin offers no difficulty; cp. plans of the imperial fora.

[25] See p. 60, n. 36.

[26] Suet. *Aug.* 57.1. See p. 60.

[27] On the worship of Apollo under Augustus, cp. Carter 164-167; Wissowa 296 f.

[28] Hor. *Epist.* 1.20.1 f.; cp. 2.3.345: Hic meret aera liber Sosiis.

[29] Varro *Ling.* 5.46; possibly Hor. *Epist.* 2.1.268-270; Prop. 4.2.2 ff.; Jordan 1.2.469.40; Gilbert 3.416; Platner 173. See p. 68, and n. 65.

[30] Liv. 1.19.2; Ov. *Fast.* 1.258; Serv. on Virg. *Aen.* 7.607; Jordan 1.3.327; Platner 191.

[31] See p. 63.

[32] *Schol.* on Hor. *Epist.* 1.20.2, 2.3.345. Little confidence can be placed on Acron's comment that they were on the rostra.

[33] According to Platner 257, there is no certain reference to an arch of Janus across the Vicus Tuscus.

Our considerations, then, have led us to infer that from the beginning of the Empire both the Argiletum and the Vicus Sandaliarius were popular book quarters. The former probably yielded precedence, because the lower end of it may have become absorbed in the Forum Palladium, but the Vicus possibly profited by the overflow from the contracted Argiletum, so that by the time of the Antonines it contained, as Galen says, most of the bookshops in Rome.[34]

Although there seems to be insufficient authority for yoking the Argiletum with the Subura as "the most crowded, noisy, and disreputable" quarter in the city,[35] the description appears fairly accurate for the Subura itself. This had evidently been a residential section at one time. Julius Caesar is said to have dwelt there in an unpretentious house, until as Pontifex Maximus he took up his abode in the Domus Publica;[36] but it had become a district *clamosa*[37] and *fervens*,[38] from which residents were moving to the Esquiline or even to distant places of peace and quiet.[39] It was possibly one of the busiest parts of Rome. At its entrance where it joined the Argiletum and apparently the Vicus Sandaliarius,[40] sat a woman barber, says Martial; no "clipper" was she, he adds, but a veritable "shaver": "Non tondet, inquam. Quid igitur facit? Radit."[41] Within its confines could be secured all the produce of market, dairy, and poultry yard: fowl, eggs, fruit, meat, olives, and vegetables—as the epigrammatist humorously describes it: "My diminutive fields bear nothing but myself, but whatever is sent to you by your Umbrian bailiff, your husbandman, your famed country estate three miles from the city, your villa in Tuscany or Tusculum, this I have raised for me all over the Subura."[42] Inscriptions locate in this quarter *crepidarii*,[43] *ferrarii*,[44] *impiliarii*,[45] *lanarii*,[46] *lintearii*,[47] *praecones*.[48] The neighborhood seems also to have been infested with brothels, and *quales*

[34] See n. 12.
[35] Platner 457.
[36] Suet. *Iul*. 46; cp. *Gramm*. 7.
[37] Mart. 12.18.2.
[38] Juv. 11.51.
[39] *Id*. 3.5; 11.51.
[40] See pp. 59 f.
[41] Mart. 2.17.
[42] *Id*. 7.31; cp. 10.94.
[43] CIL. 6.9284.
[44] *Ib*. 9399.
[45] *Ib*. 33862.
[46] *Ib*. 9491.
[47] *Ib*. 9526.
[48] *Ib*. 1953.

in media sunt Subura, as a characterization of women, was equivalent to *famae non nimium bonae.*[49]

In spite of its unenviable connections in the concrete, the phrase *in media Subura* appears to have held no unpleasant associations in the abstract, but to have signified the "heart of Rome" as vitalizing the throbbing, pulsating center of daily toil and business. Juvenal, for instance, gives expression to Hannibal's avowed ambition by the words: "Media vexillum pono Subura";[50] and Martial in a poem of tribute to Marcella, the patroness of his last years at Bilbilis, declares her worthy to grace the imperial court, and unsurpassed either by a native born daughter of Rome or a foster child of the nation's capital, proclaiming:

> Nulla nec in media certabit nata Subura
> nec Capitolini collis alumna tibi.[51]

Furthermore, the Subura seems to have continued to number the residences of prominent men within its limits. According to Martial,

> Atria sunt illic consulis alta mei:
> laurigeros habitat facundus Stella penatis.[52]

Perhaps Stella lived on or near the Clivus Suburanus which led up the slope of the Cispian Hill to the homes of other friends of the poet, such as Pliny the Younger[53] and Paulus.[54] No doubt it, as well as the Esquiline, was fast becoming a favorite residential quarter; for its muddy flags were always crowded, and if we are to believe the epigrammatist, it was as much as one's life was worth to break through the long ranks of mules as they dragged up great blocks of marble.[55]

The trading district that stretched south of the Forum to the Tiber is generally considered to have labored under a rather unsavory reputation. Horace in recounting various providers of luxuries for a prodigal gourmand, notes:

---

[49] Mart. 6.66; cp. 9.37; 11.61.1-4; 11.78.11.

[50] Juv. 10.156.

[51] Mart. 12.21, cp. 31.

[52] *Id.* 12.2(3). 10 f.

[53] *Id.* 10.20(19).3-5: "Facundo mea Plinio Thalia / i perfer: brevis est labor peractae / altum vincere tramitem Suburae."

[54] *Id.* 5.22.2, 5: "Sint mihi, Paule, tuae longius Esquiliae. / . . . alta Suburani vincenda est semita clivi." Friedländer, in his ed. of Mart., suggests that the subject of the epigram is probably Velius Paulus, cp. Friedländer-Gough 4.318.

[55] Mart. 5.22.6-8.

227   Edicit piscator uti, pomarius, auceps,
      unguentarius ac Tusci turba impia vici,
      cum scurris fartor, cum Velabro omne macellum,
      mane domum veniant.[56]

The poet's arrangement may be roughly topographical, as was Plautus's in his familiar description of life in the Forum of his day.[57] In this case, verse 227 may refer to the market section around the Forum which is not included in the succeeding lines; it may, however, allude only to the Vicus Tuscus. The phrase *turba impia*, according to the context (cp. v. 231) and to the explanation which is most acceptable to the scholiasts, presumably indicates *lenones*, and implies that the Vicus Tuscus had not altogether lost the character attributed to it by Plautus, who reported: "Ibi sunt homines qui ipsi sese venditant."[58] It is possible that the "impious rabble" includes the *fartor cum scurris*. The term *fartor* was sometimes applied to special cooks who prepared sausages or various stuffed dainties;[59] it is usually so understood here, but the combination "cook with buffoons" is perplexing. The thought seems to require a meaning for *fartor* which is more consistent with *scurra*. Now Porphyrio, commenting on verse 229, records the definition *nomenclator*, and from Festus we may add the clause, "qui clam velut infercirent nomina salutatorum in aurem candidati."[60] We may also recall that the Latin *farcio* is the source of the English word, "farce." The application of this term to a dramatic composition, according to the New English Dictionary, probably developed through the medium of old-French *farce*, which "occurs as the name for the extemporaneous amplification or 'gag,' or the interludes of *impromptu buffoonery*, which the actors in the religious dramas were accustomed to interpolate into their text." But these medieval French actors doubtless had some precedent for their use of the word in this way. It is not improbable, therefore, that *fartor* could be applied figuratively to any kind of "stuffer, and that the *fartor* whom Horace had in mind was not a "cook," but a "chief jester," or "farce actor," who, with his troop of buffoons, was engaged to amuse the host and his guests.

Although the reputation of the Vicus Tuscus suffered from the presence of its *turba impia*, the street seems to have harbored some

[56] Hor. *Sat.* 2.3.227-230.
[57] Plaut. *Curc.* 467-485.
[58] *Ib.* 482.
[59] Plaut. *Truc.* 104; Ter. *Eun.* 257; Cic. *Off.* 1.150; *schol.* on Hor. *Sat.* 2.3.229; Harcum 73 f.
[60] Fest. 88 M.

shops of the better sort. In the passage which has just been under discussion, we find that Horace connects the *unguentarius* with this quarter; and it is possibly to the same street, as a place for the purchase of perfumes and spices, that the poet alludes in the verses:

> (Ne) cum scriptore meo, capsa porrectus operta,
> deferar in vicum vendentem tus et odores
> et piper et quidquid chartis amicitur ineptis.[61]

These lines are as interesting for their inference as for their direct reference. We have previously noted that there were probably bookshops in the Vicus Tuscus.[62] Horace, seeing a glimmer of humor in the situation, utilizes it to obtain a characteristically merry surprise at the end of his Epistle. He has been making one of his apologies to Augustus for failing to apply his humble genius to the composition of a work in the Emperor's honor, which would require epic powers. He concludes that he cannot risk a bad poem, for fear that his book will be carried into the Vicus Tuscus, not to be sold by the Sosii perchance, but to be used as common wrapping paper for pepper and incense, a fate to which worthless manuscripts were sometimes consigned.[63] By Martial's day at least, the street also had shops where the finest imported silk was to be purchased.[64] As the Vicus improved in character, it apparently changed its name, to rid itself, perhaps, of its old reputation. In the time of the scholiasts, it seems to have been called the Vicus Turarius;[65] the fourth century regionary catalogue, the *Notitia*, makes no record of a Vicus Tuscus in the eighth region, but it contains the item, *vicum iugarium et unguentarium*.[66]

The Velabrum, we have noticed, was mentioned by Horace as a special market center; to Martial, *Velabrensis* recalled principally *caseus fumosus*, a very delicious cheese.[67] Its situation, as Platner remarks citing Macrobius,[68] made the district a *locus celeberrimus urbis*. A great open mart near river and Forum, it formed an important medium of traffic, and offered for sale every variety of provisions and food sup-

---

[61] Hor. *Epist.* 2.1.268-270.
[62] See p. 64.
[63] Cp. Mart. 3.2.5: "Vel turis piperisve sis cucullus." See p. 26.
[64] Mart. 11.27.11, *prima . . . de Tusco Serica vico*.
[65] Pseudo-Ascon. on Cic. *Verr.* 1.154; *schol.* on Hor. *Sat.* 2.3.228; *Id. Epist.* 1.20.1, 2.1.269; Jordan 1.2.469.40.
[66] Cp. Jordan 2.553.
[67] Mart. 11.52.10; 13.32.
[68] Macr. *Sat.* 1.10.15; Platner 394.

plies.⁶⁹ Other market centers (and the commercial prosperity of the Empire supported several of them⁷⁰) were doubtless less congested and more reputable and sanitary. Produce dealers received substantial encouragement from the emperors themselves; for the *macellum* that had been built behind the Basilica Aemilia by M. Fulvius Nobilior in 179 B. C.,⁷¹ had been supplemented by the Macellum Liviae of Augustus on the Esquiline,⁷² and by the Macellum Magnum of Nero on the Caelian.⁷³ Those who stocked them furnished the city not only with necessities but with every luxury of field, stream, and forest, but the majority of dealers demanded such high prices⁷⁴ that Martial found excuse for forsaking Rome, because "here it costs to go hungry and marketing is rioting."⁷⁵

The quarter beyond the Tiber, to which were relegated all industries likely to become a nuisance, such as tanneries and sulphur plants, has previously been discussed.⁷⁶ Its formerly obnoxious features grew gradually less offensive, it seems, under what Juvenal deemed the slogan of the Empire: "Lucri bonus est odor ex re qualibet."⁷⁷

Inscriptions give evidence that the Sacra Via was a fashionable shopping street in the days of the Empire. There are references to florists, fruitiers, and especially cutters of precious stones, dealers in pearls, bronze-chasers, goldsmiths, and jewellers.⁷⁸ In Martial's time, an uptown district also was in favor with the "Four Hundred." This was the Saepta, the enclosure in the Campus Martius which had originally served as the voting place of the *comitia centuriata*; but later, with a succession of arcades flanking the Via Flaminia,⁷⁹ it offered a veritable Rue de Rivoli, "where golden Rome displayed her riches."⁸⁰ Martial peoples it for us with such vividness that shadowy forms assume definite

---

⁶⁹ Cp. CIL. 6.467, 9184, 9259, 9671, 9993.
⁷⁰ Hor. *Sat.* 2.3.229; Mart. 10.59.3; Juv. 11.64.
⁷¹ Jordan 1.2.432-436; Gilbert 3.207-209; Platner 274, 460.
⁷² Gilbert 3.237 f.; Jordan 1.3.344 f.; Platner 274, 470.
⁷³ Gilbert 3.238; Jordan 1.3.237 f.; Platner 441.
⁷⁴ Hor. *Sat.* 2.3.225 ff.; 2.4.76; *Epist.* 1.15.31; Mart. 10.37.19; 10.59.3; 12.62.9 f.; 13.85.1; Juv. 5.95; 11.9-11.
⁷⁵ Mart. 10.96.9.
⁷⁶ See p. 19.
⁷⁷ Juv. 14.204 f. Cp. Hor. *Epist.* 1.1.65 f.; Sen. *Epist.* 115.14; Juv. 14.205-207; Suet. *Vesp.* 23.2 f.
⁷⁸ CIL. 6.9207, 9212, 9221, 9214, 9239, 9283, 9418 f., 9545-9549, 9795, 9935; cp. *Notitia* (Jordan 2.553), *Porticum margaritarium*; Platner 315 f. See p. 8.
⁷⁹ Gilbert 3.174 f.; Jordan 1.3.558-562; Platner 345 f., 384-386.
⁸⁰ Mart. 9.59.2.

shape and the place becomes a reality of actual life. We see the obsequious client pursuing the man of wealth,[81] and the foppish pretender promenading as a gay, young knight, although his last cent has probably been spent on his attire and he has pawned his ring to buy a cheap dinner.[82] Next we catch sight of an envious slave or freedman tearfully heaving a sigh from the bottom of his heart, because he cannot buy the whole Saepta and take it home with him.[83] Finally, we note the familiar counterpart of an unscrupulous modern shopper: he inspects the fairest slaves, those that are not exposed for public sale but are shown in private only to the élite; he examines rare ivory ornaments, and furniture of choicest wood inlaid with tortoise shell; he tests Corinthian bronzes, criticizes statues of Polycleitus, detects flaws in the finest crystal (yet sets aside a few specimens for future consideration); then he scrutinizes goblets of Mentor's chasing, seeks out the richest and most costly precious stones, and finally after a whole day's fatiguing efforts, purchases a couple of common cups for a two cent piece, and—carries them home himself![84]

The *tabernarius* appearing most frequently in Martial's pages, due to the repeated reference to one individual, is the *unguentarius*. Although there were some perfumers whose names one might hesitate to reveal,[85] they seem in general to have kept shops of the better kind, and to have suffered very little from disdain or disapproval.[86] According to Trimalchio's facetious exposition of astrology, they were born under Libra.[87] Considering the lavish use of ointments and essences, not only for the person,[88] but for banquets[89] and for spraying at the theater,[90] the demand for *unguentarii* must have been unusually great. Martial mentions Niceros as one who dealt in rich oils and scented unguents in his time.[91]

[81] *Id.* 2.14.5.
[82] *Id.* 2.57.
[83] *Id.* 10.80.
[84] *Id.* 9.59.
[85] Cp. Juv. 2.40-42. Petron. 74 mentions an *unguentarius* who was a slave attendant of a wealthy woman.
[86] Hor. *Sat.* 2.3.228, see pp. 67 f.; Juv. 14.203 f., see pp. 18 f.
[87] Petron. 39.
[88] *Id.* 47, 77 f.; Mart. 1.87.2; 2.29.5; 3.12.4 f.; 3.55; 7.41; 11.49(50).6; 12.55.7; 12.65.4; 14.59.146; Juv. 2.41.
[89] Hor. *Carm.* 2.7.22 f. *et passim*; *Sat.* 2.3.228; Petron. 60, 68, 70; Mart. 3.12; 3.82.26-28; 11.15.6; Juv. 8.86.
[90] Mart. 5.25.7 f.; 8.33.3 f.
[91] *Id.* 6.55.3; 10.38.8; 12.65.4.

But the best known perfumer of that period appears to have been Cosmus, to whom allusion is made in about fifteen epigrams. Since Juvenal speaks of the same man,[92] the appellative, although it suggests the business of its bearer, is probably not one of Martial's fabrications.[93] Cosmus may have been a freedman who adopted an appropriate name when he set up in trade; of course the appropriateness may have been purely accidental—one of those amusing coincidences which are frequently found in every generation, such as Rose, the florist. Giese, who has made a special study *de personis a Martiale commemoratis*, singles out only *Cosme* of 4.53 as imaginary,[94] but it seems logical and legitimate to take this epigram in connection with the others, and thus obtain a strong contrast between the fastidious customers suggested in the lines and the slovenly Cynic "cur" described. Cosmus presumably did not make a specialty of common *glaucina*[95] and *capillare*,[96] or *opobalsama*, a favorite with men,[97] but carried a stock of choicest unguents, *cinnamum*,[98] *pastilli*,[99] and catered particularly to feminine tastes.[100] Certain specialties bore his name: Martial terms a rich perfume *Cosmianus*,[101] and dubs spikenard *folium Cosmi*.[102] Because of the lingering aroma, Martial suggests an *ampulla Cosmiana* as a welcome drinking flask for one who thirsts for nard wine.[103]

Before leaving the subject of *unguentarii*, we should note briefly two disputed passages which may have reference to specific individuals. The first is Martial 7.41:

> Cosmicos esse tibi, Semproni Tucca, videris:
> cosmica, Semproni, tam mala quam bona sunt.

The question pertains to the initial word in each verse, for which Harper's lexicon gives the signification "cosmopolite." This affords a plausible

---

[92] Juv. 8.86. Nicolaus Perottus, *Cornucopia* 200.26 (Aldine ed., 1513), quotes this and also the following fragment from Petronius: "Affer nobis, inquit, alabastrum Cosmiani."

[93] Stephenson, on Mart. 3.55.1, inclines to the view that it is fictitious.

[94] Giese 12.

[95] Mart. 9.26.2.

[96] *Id*. 3.82.28.

[97] *Id*. 14.59. Cp. Juv. 2.41.

[98] Mart. 3.55. Cp. 6.55.1.

[99] *Id*. 1.87.2.

[100] *Id*. 11.49(50).6; 12.55.7; 12.65.4; 14.59.2.

[101] *Id*. 11.15.6; 12.55.7. See n. 92.

[102] Mart. 11.18.9; 14.146; cp. 11.27.9; 14.110.2.

[103] *Id*. 14.110; cp. 3.82.26; Juv. 8.86; Marquardt 2.461,650.

rendering, but after considering the connotation of *Cosmus* and *Cosmianus* in the epigrams, we are inclined to agree with those who find more point in the interpretation: "You think, Sempronius Tucca, that you are well scented. Scents, Sempronius, are bad as well as good!"[104] In the second dubious allusion, "cuius olet toto pinguis coma Marcelliano,"[105] Giese finds reference to a *Marcellus, myropola*.[106] However, a glance at the opening line of the epigram, "Rufe, vides illum subsellia prima terentem," and at such phrases as *tota . . . Subura*,[107] *urbe . . . tota*,[108] and *Marcelli Pompeianumque*,[109] *Pompeiano vela negata Noto*,[110] and *scaena Marcelliani theatri*,[111] leaves little doubt that the ablative is a locative with *theatro* to be supplied.

In determining the reputation of *tabernae* and their managers in general, we must beware of specializations. For instance, Horace's advice to writers of *saturae*:

> Ne quicumque deus, quicumque adhibebitur heros,
> regali conspectus in auro nuper et ostro,
> migret in obscuras humili sermone tabernas,[112]

does not, we believe, contain a thrust at all shops. The adjective, *obscuras*, whch would be redundant if *humili sermone* were broad in its application, possibly has a counterpart in *arcana* of a passage from Martial:

> Et blando male proditus fritillo,
> arcana modo raptus e popina,
> aedilem rogat udus aleator.[113]

It is doubtless descriptive therefore of such places as taverns[114] or especially cookshops.[115] The epigrammatist would certainly disclaim *humilis sermo* as a disparaging generalization; for in prophesying the spread of fame and glory for a book of poems which he is sending to Cassius Sabinus, he remarks:

---

[104] Cp. Mart. 1.87; 2.12.
[105] Mart. 2.29.5.
[106] Giese 22.
[107] Mart. 7.31.12.
[108] *Id.* 1.2.5 f.
[109] *Id.* 10.51.11.
[110] *Id.* 11.21.6.
[111] Suet. *Vesp.* 19.1.
[112] Hor. *Epist.* 2.3.227-229.
[113] Mart. 5.84.3-5.
[114] In Petron. 80, *humilis taberna* refers to a lodging house. See pp. 11-13.
[115] See pp. 15-18.

> Te convivia, te forum sonabit,
> aedes, compita, porticus, tabernae.[116]

The barbers', the perfumers', and the booksellers' were among the favorite gathering places in Rome.[117]

From a financial standpoint shops "paid." Juvenal cites the case of a freedman under Domitian, who had acquired a knight's fortune because "quinque tabernae quadringenta parant."[118] His prosperity scandalized the aristocracy, but it gained him admittance at the doors of the mighty, and showed the trend of the times. Although the domestic life of the small tradesman was no doubt often confined to second floor lodgings[119] which communicated with the *taberna*, but which with it were separated from the rest of the house, this state of affairs appears not to have been so common or so inevitable as Typaldo-Bassia would lead us to infer.[120] *Tabernae* at Pompeii, at least, permit some interesting conclusions. As Mau-Kelsey point out, they formed the outer parts of houses fronting on the principal thoroughfares.[121] A number of them had no connection with the private apartments and were doubtless rented; many, however, opened upon the *fauces* or inner rooms and were presumably the house-owner's or tenant's place of business. The latter conditions prevailed, not only in small and ordinary homes, but in some of the finest in the town. To note a few instances, it was the case in the House of the Tragic Poet, which Mau-Kelsey designate as "among the most attractive in the city";[122] it was also true of the House of the Faun, accounted "among the largest and most elegant in Pompeii," with mosaic floors which "are the most beautiful that have survived to modern times."[123] Such an environment for shops whether private or rented, in marked contrast to Cicero's declaration that *tabernarii* were rioters and the dregs of the populace,[124] betokens their reputability and

---

[116] Mart. 7.97.11 f. Cp. Petron. 140.

[117] Plaut. *Amph.* 1011-1013; *Epid.* 198 f.; Ter. *Phorm.* 89 ff.; Hor. *Sat.* 1.7.1-3; Gell. 18.4.1; cp. Lysias *Orat.* 24.20; Demos. *vs. Phorm.* 13. See pp. 59, 90.

[118] Juv. 1.105 f. It is scarcely possible to say whether the *tabernae* in question were a property or business investment or both.

[119] Cp. CIL. 4.138, 1136.

[120] Typaldo-Bassia 24.

[121] Mau-Kelsey 276 ff.

[122] *Id.* 313 ff.

[123] *Id.* 288 ff. Cp. the House of Pansa, a whole *insula* with shops both connected with the dwelling and otherwise (349 ff.). The House of the Vettii is well known for its Cupid and Psyche pictures; "prosaic daily toil has nowhere been more happily idealized" (331 ff.).

[124] Cic. *Dom.* 13; *Flacc.* 18.

orderliness, and makes possible the inference that under the Empire, in municipal towns at least, shopkeepers included men of means and position. At Rome they naturally did not move in exclusive circles, and were not always known personally to men of high degree; but they were apparently people of influence in their own communities. Under Augustus we find some of them wielding a political power not to be ignored. In fact, an aspiring candidate found it advisable to secure an informant's services and pay deference to the man "behind the counter." Says Horace,

> Mercemus servum qui dictet nomina, laevum
> qui fodicet latus et cogat trans pondera dextram
> porrigere: 'hic multum in Fabia valet, ille Velina;
> cui libet hic fascis dabit eripietque curule
> cui volet importunus ebur.'[125]

We can not but surmise that there may have been cases where continued political influence, added to financial success in business, brought social recognition even from the nobility without a nomenclator's aid.

XXIV

Textores

Spinning and weaving[1] were of course among the primitive occupations at Rome, but they do not appear to have been represented in the early *collegia* listed by Plutarch.[2] This was doubtless because they constituted merely household tasks. By the time of Plautus, however, there were independent dealers in woollen goods and woven fabrics;[3] and the Andrian woman of Terence's play is said to have made her living at the loom when she first came to the Roman Capital.[4] Under the Empire there were public and private weaving-rooms, *textrinae*, which employed many hands.[5] Those who spun the thread were termed

---

[125] Hor. *Epist.* 1.6.50-54. I have adopted what seems to me to be the most obvious interpretation of *trans pondera* (51), although this is contrary to Acron, who says that *pondera* means "stepping-stones." Other suggestions are "beyond our balance" and "over the weighted tassels of our gowns." Cp. Wickham's and Wilkin's notes.

[1] Cp. Ov. *Met.* 6.53-69; Marquardt 2.517-527; Blümner, *Tech.*, 1.120-170.

[2] See p. 1.

[3] Plaut. *Aul.* 508 ff.

[4] Ter. *Andr.* 74 f.

[5] Typaldo-Bassia 8; Blümner, *Tech.*, 1.166.

lanificae;[6] weavers in linen were sometimes called *linteones*[7] or *lintearii*[8] to distinguish them from workers in wool, *lanarii*;[9] and inscriptions record certain specialists such as carders, *carminatores*,[10] *pectinarii*.[11] Among the papyri discovered at Oxyrhynchus is one of the time of Augustus containing a Latin account of wages paid to *textores, conductei,* and a *magister*; according to this, the weavers received three and one half asses per day; the hirelings, four asses; and the overseer, six asses.[12] Wages at Rome were possibly higher.[13]

The satirists do not throw much light upon the relations of spinners and weavers to society. In many homes garments were still woven by slaves under the direction of the *domina, lanipendia,* or *vilica*.[14] Horace uses the noun *textor* quite impersonally as an ablative of means in the following rhetorical question:

> Quid si quis vultu torvo ferus et pede nudo
> exiguaeque togae simulet textore Catonem,
> virtutemne repraesentet moresque Catonis?[15]

The juxtaposition of *textore* and *Catonem* is noticeable; it is doubtless an intentional oxymoron.

In denouncing the custom of osculatory greeting which had been prevalent at Rome since the beginning of the Empire,[16] Martial warns the returned traveler that from the omnipresent throng,

> Hinc instat tibi textor, inde fullo,
> hinc sutor modo pelle basiata.[17]

He groups these with some of the most loathsome and objectionable characters, but he apparently does not mean to place them on the same level. His purpose is to set forth the universality of the kissing habit; in order to include the whole mass of the people he names the rough,

---

[6] Many references in the satirists to spinners designate the *Parcae*: Mart. 4.54. 5-10; 6.58.7 f.; Juv. 12.64-66; cp. Mart. 4.73.3 f.; 6.3.5 f., on Julia, daughter of Titus, as spinner or one of the Fates; 7.96.3 f.; 9.76.6 f.; 10.44.5 f.; 10.53.3; Juv.3.27.

[7] Plaut. *Aul.* 512; Serv. on Virg. *Aen.* 7.14.

[8] CIL. 6.9526.

[9] Plaut. *Aul.* 508; CIL. 5.4501; 6.9491; 11.1031.

[10] CIL. 11.1031.

[11] CIL. 5.4501.

[12] Grenfell and Hunt, *Oxyrhynchus Papryi*, 6.233.737 (London, 1904).

[13] Cp. West 294 f., 304. West's discussion, 293 ff., is interesting as a caution against comparing ancient wages with modern without making all due allowances.

[14] Juv. 2.57 and *schol.*; 6.476 and *schol.*; 11.69; Marquardt 1.156.

[15] Hor. *Epist.* 1.19.12-14.

[16] See pp. 22, n. 13; 58.

[17] Mart. 12.59.6 f.

unkempt, country peasant, the class engaged in trade, the dregs of the populace. He possibly gave no more thought to the selection which he made to represent the second division than is taken nowadays for the stock phrase, "butcher, baker, and candlestickmaker."[18]

Juvenal presumably considered the carder of wool on a far lower plane than the *grammaticus*, and he deemed it quite unjust that the latter had to be at his desk at an early morning hour,

> Qua nemo faber, qua nemo sederet
> qui docet obliquo lanam deducere ferro.[19]

Perhaps the second verse refers to the foreman of a *textrina*. Since the order is probably intended to secure a climax, we may concede that the poet would rank *pectinarii* below *fabri*. In his bitter attack against certain hypocritical philosophers, Juvenal expresses utter scorn for men who were taking up the art of weaving, which he considers not only effeminate, but the task of a slave, "horrida quale facit residens in codice paelex."[20] In another Satire, he contrasts the glory of a lineage traced from royal blood to descent from one "quae ventoso conducta sub aggere texit."[21] The antithesis may not be the satirist's own; for it is offered in stinging irony as the words of a worthless nobleman, who, as Juvenal intimates, might easily have been surpassed by one less highly connected. The verses therefore may from the poet's point of view merely signify humble rank. On the other hand, the *agger* mentioned is no doubt the Tullian Embankment that crossed the plateau between the Porta Collina and the Porta Esquilina,[22] a quarter which does not seem to have been in good repute. The section near the Esquiline was possibly improved as a promenade[23] when the Horti Maecenatis[24] were laid out, but it had previously been in the vicinity of the Potter's Field.[25] At the Colline Gate were the barracks of the Praetorian Camp.[26] Juvenal refers to the Rampart elsewhere as the resort of idlers and low characters who were

---

[18] The play on words that *sutor* offered may have tempted him to choose providers of wearing apparel, or he may have had personal motives, as is suggested on p. 22.

[19] Juv. 7.223 f. Cp. CIL. 5.4501, LANARI. PECTINAR / SODALES—from Brescia; 11.1031, LANARIORVM / CARMINATOR—from Brescello.

[20] Juv. 2.54-57 and *schol.* Note the masculine endings of nouns in the inscriptions cited.

[21] Juv. 8.43.

[22] See p. 12.

[23] Hor. *Sat.* 1.8.14-16.

[24] Gilbert 3.361 f.; Jordan 1.3.346 f.; Platner 464-466.

[25] Hor. *Sat.* 1.8.6-13; Jordan 1.3.261, 265-270; Platner 445 f.

[26] Jordan 1.3.385-393; cp. Juv. 16.26.

attracted by fortune tellers[27] and performing animals.[28] Weavers who lived or worked[29] in this neighborhood, therefore, could scarcely have been much esteemed.

It is logical to suppose that the standing of *textores* in general was lowered by the number of slaves both domestic and public that crowded their ranks. Moreover, since the Romans imported much woven material from Transalpine Gaul and Egypt,[30] we may infer that many foreigners of those nationalities emigrated to the Capital and engaged in the trade for which their countries were noted. Waltzing records no evidence for corporations of weavers at Rome, but inscriptions of *collegia* have been found at Brescia, Brescello, Thyateira, and Ephesus.[31]

XXV

Tignarii

Collegia Fabrum Centonariorum Dendrophorum

Plutarch lists *tignarii* among the early corporations which he ascribes to Numa.[1] They no doubt comprised originally all workers in wood, but as division of labor increased, they became subdivided into specialized groups such as *subsellarii, lectarii*.[2] It may have been one of the last named that made Horace's *Archiacis . . . lectis*;[3] for according to Porphyrio, "Archias breves lectos fecit."[4] Some apparently did not

---

[27] Juv. 6.588-591. Cp. the witches in Hor. *Sat.* 1.8.17 ff.

[28] It is to these presumably that Juv. has reference in 5.153-155. Cp. *schol.* and Mart. 14.128, 202.1.

[29] Wilson on Juv. 8.43 compares Shakesp., *Twelfth Night*, 2.4.44, "The spinsters and the knitters in the sun." It is hardly possible to decide whether the reference in Juv. is to women working in their own homes or as "hands" in *textrinae*. Inscriptions locate weavers in the Subura, see nn. 8, 9, and pp. 65 f.

[30] Cp. Mart. 4.19; 14.128, 150, 159 f.; Juv. 7.221; 9.28-30; Blümner, *Thätigkeit*, 10 f., 137 ff., 142 ff.

[31] Waltzing 2.153; 4.95.

[1] See p. 1.

[2] Cp. Kühn's register of inscriptions 37-39.

[3] Hor. *Epist.* 1.5.1 and *schol.*

[4] Cp. Hor. *Sat.* 2.7.95, *Pausiaca . . . tabella*; also "Windsor chairs." H. E. Eve, however, offers a different explanation in the Class. Rev. 19(1905).59, which is very ingenious and tempting. He connects the phrase with Plutarch's story (Life of Pelop. 10) of Archias, Governor of Thebes, who, while banqueting, received from his namesake in Athens a letter containing details of the conspiracy of Pelopidas, but cast it aside with the words εἰς αὔριον τὰ σπουδαῖα which afterward became proverbial.

adopt a speciality but retaining their versatility, manufactured anything from a bench to a Priapus.[5]

After the term *tignarii* had become restricted, it seems to have been applied regularly to carpenters and builders,[6] and to have kept its signification even after wood ceased to be the chief building material. Proof for this is offered by Gaius, who says: "Fabros tignarios dicimus non eos dumtaxat, qui tigna dolarent, sed omnes, qui aedificarent."[7] Wherever, therefore, *faber* is used without a distinguishing adjective, but in connection with building, it is evident that *tignarius* is to be supplied.[8]

Fabri Some authorities support the theory that the word *fabri*, when unmodified, was employed by writers to mean *tignarii*, and that there was little difference, therefore, between the *collegium fabrum tignariorum* or *tignuariorum* and the *collegium fabrum* of municipal towns and colonies.[9] The reasoning appears illogical; for the context may be as restrictive as a limiting adjective, as for example in Horace's expression, *marmoris aut eboris fabros aut aeris*.[10] The use of *aedificare* or kindred words, therefore, would particularize *fabri*, so that it could not accurately be said to be "unmodified." On the other hand, where neither context nor adjective specifically distinguishes the word, it seems to be collective in force, including, as the verse just quoted from Horace suggests, various workers in hard materials. Varro, for instance, says that husbandmen used to call in neighboring *medicos, fullones, fabros* rather than keep them on their farms;[11] he surely means the makers of implements as well as builders. Again, Asconius, commenting upon the *senatus consultum* against corporations, declares that *collegia* which were of advantage to the state, *sicut fabrorum fictorumque*, were allowed to remain;[12] and certainly others besides carpenters would be considered an *utilitas civitatis*. Horace, too, in the clause *tractant fabrilia fabri*[13] was in all probability thinking as much

[5] Hor. *Sat.* 1.8.1-3.
[6] Cp. Cic. *Brut.* 257.
[7] Dig. 50.16.235.
[8] Cp. Cato *Agr.* 14.1; Cic. *Verr.* 5.48; *Epist.* 9.2.5; Hor. *Epist.* 1.1.83-87.
[9] Waltzing 2.117 ff., 149; Kornemann 4.1.394; 6.2.1906; Kühn 27-28, 36.
[10] Hor. *Epist.* 2.1.96.
[11] Varro *Rust.* 1.16.4.
[12] Ascon. *Corn.* 75.67 (Clark).

[13] Hor. *Epist.* 2.1.116. Cp. Juv. 14.115 f.: "Egregium populus putat adquirendi / artificem; quippe his crescunt patrimonia fabris." The lines refer metaphorically to a miser, and offer a play on the word *fabris* in its broad meaning "artificers," and

of the skill of *aurifices, argentarii,* and *aerarii* as of that of *tignarii*. Finally, there are Martial's phrases, *faba fabrorum* and *fabrorum prandia, betae*,[14] in which *fabrorum* is evidently equivalent to *plebeia* as is indicated by Persius's corolary, *plebeia . . . beta*.[15] The metonomy is more natural if *fabri* be a generic term; for as such it would doubtless have included a very large part of the industrial population, especially since *faber* was not always used accurately, but by an extention of meaning might be loosely applied to workers in soft materials, *fictores*.[16]

In the face of this it is somewhat inconsistent to interpret *collegium fabrum* as a body of one specific group of men. The organization, it appears, acted as a fire department in municipal towns and colonies and had its members specially appointed by the government.[17] Kühn suggests that perhaps *fabri tignarii* in the *municipia* became volunteer firemen, and later in larger towns were publicly assigned this duty, being then distinguished from the strictly industrial corporations by the simple term *fabri*.[18] Now *a priori* it seems rather incongruous that of all people carpenters and builders should constitute a fire department. In the first place, considering the enthusiasm for building under the Empire, both on the part of the state and of individuals,[19] they must have been far too busy for burdensome outside duties, or too much in demand to permit their number to be considerably curtailed. In the second place, if a large portion of society was as base and dishonorable as the satirists would have us believe,[20] it might reasonably be suspected that a fire company composed exclusively of *tignarii* might be open to the temptation of increasing business for its members along both lines!

Collegia Fabrum

On the other hand, the evidence cited above suggests that a *collegium fabrum* would consist of various kinds of artificers; Pliny's lan-

---

its narrower sense of *argentarii*; cp. App. Claud. *Carm.* frg. 36: "Est unus quisque faber ipse fortunae suae."

[14] Mart. 10.48.16; 13.13.1.

[15] Pers. 3.114.

[16] Cp. Petron. 51: "Fuit . . . faber qui fecit phialam vitream"; Juv. 1.54, *fabrumque volantem=Icarum*; Blümner, *Tech.*, 2.166.

[17] Plin. *Epist.* 10.33(42), 34(43); *Paneg.* 54; Hirschfeld 239 ff.; Liebenam 104 f.; Kornemann 4.1.394 f.; 6.2.1905 f.

[18] Kühn 35 f.

[19] Hor. *Epist.* 1.1.83-87; Mart. 9.22.16; 9.46; Juv. 1.94; 14.86-91; Suet. *Iul.* 44; *Aug.* 29; *Cal.* 21; *Claud.* 20; *Nero* 31; *Vesp.* 9.1; *Dom.* 5.

[20] Cp. Petron. 116, 119; Mart. 3.38; 4.5; *et passim*; Juv. 3.21-57, 109-125; 13.23-30; 14.123-178; *et passim*.

guage, "Ego attendam ne quis nisi faber recipiatur,"[21] points in the same direction. Furthermore, although *tignarii* may well have played an important part in helping to extinguish fires, yet the construction, repair, and skilful manipulation of the *siphones, hamae, reliqua instrumenta ad incendia compescenda*[22] noted by Pliny were doubtless in the hands of several classes of specialists. Inscriptions make frequent reference to corporations of *centonarii* and *dendrophori* in connection with *fabri*.[23] There appears to be little doubt therefore that *collegia fabrum* were heterogeneous organizations.

Centonarii

The *centonarii* and *dendrophori*, of whom mention has just been made, present some difficulty. The former were presumably, as Kornemann and others maintain, manufacturers and perhaps merchants of coarse canvas cloth which was pieced together to make *centones*.[24] This term seems to have signified primarily and specifically coverings made of pieces of canvas, which were employed in warfare as a protection against missiles. Caesar refers to their use for military engines,[25] and also tells of an occasion when "almost all of the soldiers made tunics or covers out of felt or canvas or leather" (*ex coactis aut ex centonibus aut ex coriis*).[26] According to an anecdote related by Ammianus Marcellinus, some parched soldiers made a rope out of strips of linen, and drew water from a well by attaching their improvised rope to a canvas cap (*centonem*), which one of their number had been wearing under his helmet.[27] *Centones* must also have been effective against fire; for they are mentioned among the requisites for fighting a fire;[28] and in inscriptions, *collegia centonariorum* are found almost exclusively in connection with incorporated *fabri* and *dendrophori*.[29] Perhaps *centones* were employed in some way to smother flames; but judging from their use in warfare, they doubtless served as a protective covering both for apparatus and for the fire fighters themselves. References given above, for instance, permit us in imagination

---

[21] Plin. *Epist.* 10.33(42).3.

[22] *Ib.* § 2. Petron. 78 mentions *secures*, and the Digest, *centones, siphones, perticae, scalae, phormiones, spongiae, hamae, scopae*; see n. 49.

[23] See a catalogue of them in Waltzing 4.50-72, 76-80; Kornemann 6.2.1907-1911; Kühn 33-35, 39.

[24] Waltzing 2.146; Kornemann 4.1.395; Blümner, *Tech.*, 1.209.

[25] Caes. *Civ.* 2.9.

[26] *Ib.* 3.44.

[27] Amm. 19.8.8.

[28] See n. 22.

[29] Cp. register, Waltzing 4.76 f.; Kornemann 6.2.1907-1911.

to equip the fire brigade in a special costume of canvas cap and tunic.

Not only did the military custom which we have described pass into a public, civic institution, but as in the case of the *sagum*,[30] it was also taken up in private life, especially by slaves, workmen, and the poorer classes. Consequently we find the term *cento* applied to the cheap garb of common people, to patchwork curtains, and to bed coverlets.[31] It was no doubt loosely used for a number of things which had some similarity to military *centones* in appearance or in structure, even if they were not made of canvas; like *vestis*, the word appears to have been applicable to any cover, whether a garment or a spread. It is easy to see how the meaning "patchwork" developed; but the rendering "rag-covering" seems sometimes to be carried a step too far, especially when it is taken to lay special emphasis upon the *age* or *worthlessness* of a thing, instead of upon the fact of its being *pieced* together. A "patchwork quilt," for instance is made not of rags, but of good *pieces*. Encolpius in Petronius admits that an old tunic which he and Ascyltos were claiming was a bunch of rags not fit for good *centones*.[32] In Plautus the word appears as a colloquialism for the "patched up tale" of a boastful soldier.[33] In late Latin it was assigned to a poem which was a compilation of various verses and parts of verses from another poem;[34] it was a classic, however, not a worthless work, that was usually chosen to be cut up and pieced together according to the *mos centonarius*; and so there were *Homerocentones* and *Vergiliocentones*.[35] The thirteenth Idyll of Ausonius, for instance, known as the *Cento Nuptialis*, was based upon Virgil.[36] The phrase *vestiarius centonarius* is found in Orelli 4296 and is cited by the lexica and some authors, but Orelli's inscription is classed among the *Falsae*, CIL.5.30. There is apparently no good reason, therefore, why *Echion centonarius* in Petronius 45 f. should be considered a "rag dealer" or an "old clothes dealer," as he is generally regarded.[37] Had Echion originally been such a lowly personage, he would doubtless have changed his occupation to accord with his higher

[30] Cp. Blümner, Müller's *Handbuch* 4.2.216 f.

[31] Cato *Agr*. 2.3; 10.5; Lucil. in Non. 176.1 M; Petron. 7, 14; Juv. 6.121 and *schol*.; Fest. 237 M.

[32] Petron. 14.

[33] Plaut. *Epid*. 455.

[34] Teuffel-Schwabe-Warr 1. § 26.

[35] Tert. *Praescr*. 39; Hieron. *Epist*. 103.7; Isid. *Orig*. 1.39.25.

[36] Teuffel-Schwabe-Warr 2. p. 367.

[37] Cp. Peck 107 (New York 1898) and Lowe 61 (London 1905), trans. of the *Cena Trimalchionis*, and Heseltine 75 (London, 1913), ed. and trans. of Petron.

station, following the example of Phileros, who had once been a peddler with things to sell on his back,[38] but had later become an advocate, and a match for the best of them. Note the chatter of Echion: he criticises past shows, gives advance information about those that are to come, and indulges in personalities about their doners. His children, he prophesies, are destined for "careers"; one, who is not especially interested in books, is to learn the business of a barber, an auctioneer, or an advocate; but the other is a student of Latin and Greek and is in a fair way to become a *grammaticus*. Surely the father is more than a "rag dealer"; interpreted as a prosperous canvas manufacturer, perchance even a member of the local fire department, this character assumes grander proportions, and we can well understand his optimism, his pride in his country and his boys, and his effervescent sense of importance.

For the identification of the *dendrophori* no satisfactory explanation has as yet been advanced. It is perfectly clear from inscriptions that there was a religious brotherhood of this name, concerned with the cult of Cybele and Attis, at whose festivals they carried branches of trees in procession.[39]

Dendrophori

They seem also to have venerated Silvanus,[40] and to have been connected with the cult of the emperors.[41] Godefroy believed that there were two distinct colleges, one a religious body of priests consecrated to the worship of Cybele, the other a civil and industrial corporation composed perhaps of carpenters or *negotiatores* in timber.[42] Waltzing considers the two institutions one and the same, and is inclined to the belief that their personnel consisted of lumber merchants, who originally under a native Latin name, perhaps *lignarii*, had supplied lumber for public use and had therefore been detailed by Claudius, when he instituted the great April festival in honor of Magna Mater, to secure the sacred pines for the celebration. They had subsequently dedicated themselves to the worship of the goddess and had been appointed by the state to look after her rites. Gradualy the name *dendrophori*, which had at first indicated only their religious function[43] in a cult in which all nomen-

[38] Cp. Petron. 38.

[39] Cumont, *Oriental Religions in Roman Paganism*, 56-58 (Chicago. 1911); Taylor, *Cults of Ostia*, 64 f. (Bryn Mawr Dissertation, 1912).

[40] Von Domaszewski, *Silvanus auf lateinischen Inschriften*, *Philol*. 61(1902).15; Cumont, *Real-Encycl*. 5.1.218; Taylor, *op. cit.*, 40.

[41] CIL. 5.3312, 5275; 9.3938; 13.5153; Cumont, *l. c.*

[42] Godefroy on *Cod. Theod.* 14.8.1; 16.10.20.

[43] Cp. *cannophori* of Cybele, *cistophori* of Bellona, *pastophori* of Isis: Darem.-Saglio 1.686; Cumont, *Oriental Religions*, 56, 94.

clature was Greek, was extended to their trade and caused their Latin name to fall into disuse.[44]

There are at least two strong objections to this theory. In the first place, as Waltzing himself confesses, it would show the college to have been a rare exception, in that, though of private character itself, it maintained a public cult;[45] in the second place, it does not sufficiently explain its relation to *fabri* and *centonarii*. Possibly the association arose in this fashion: Early in the Empire on the analogy of the *cohortes vigilum*, established at Rome by Augustus in 6. A. D. as a fire and police department,[46] *collegia fabrum* were appointed by the government in municipal towns and colonies. It is natural that their membership should have been made up from these particular ranks of the industrial orders, since it was from them, no doubt, that volunteers had previously rushed to conflagrations, taking whatever implements they had to offer. It is quite possible, too, that in forming the new state departments, men were chosen in some cases who already belonged to special guilds;[47] for instance, a certain T. Flavius Hilario is seen to have been DECVR. COLL.FABR. . . MAG.QVINQ.COLL.FABR.TIGNARIOR.[48] Now mechanics would naturally know and avail themselves of the military device of using *centones* to protect engines. It was doubtless found that these were also advantageous for smothering flames, or as we have previously suggested, that they could be converted into serviceable apparel for firemen; and so, to increase the efficiency of fire companies, *centonarii*, who were possibly already largely engaged in the service of the state as the manufacturers of a commodity of warfare, were annexed to the *collegia fabrum* or associated with them.

[44] Waltzing 1.249-251; 2.123 f., 148. Cp. Paris in Darem.-Saglio 2.101, and fig. 2330; Kornemann 4.1.396; Cumont, *Real-Encycl.* 5.1.216. In his *Oriental Religions*, 58, Cumont suggests that the *dendrophori* were wood cutters and carpenters, able to fell the divine tree of Attis, and also to cut down the timbers of burning buildings.

[45] Waltzing 1.253.

[46] Mommsen, *Staatsverw.*, 2.484-487 (Leipzig, 1887); Abbott, *Roman Political Institutions*, 281 (Boston, 1911).

[47] Cp. Waltzing 1.351 ff., "Nous rencontrons beaucoup d'hommes affiliés à deux ou plusieurs collèges à la fois," etc.

[48] CIL. 14.2630. Cp. 9.5450, where MAG.COLLEG / FABR. is also MAG.ET.Q.SODAL / FVLLONVM; this inscription permits the inference that in small towns at least, *collegia fabrum* admitted others than *fabri*.

Meanwhile the need of special fire-fighting implements would be increasing. Long poles, *perticae*,[49] were found to be efficacious in an emergency; and ladders, *scalae*,[49] became essential to cope with the situation in the case of high buildings. As methods improved and inventions increased, mechanical devices probably became larger and heavier, and there was need of a greater force of men to transport them. It would also be expedient to have special workers to remove furniture from burning buildings and carry it away to safety. There was need, then, of a Company of Porters to render the fire department thoroughly efficient, and there were state *collegia* ready at hand to provide them; for, as we have noted, since the time of Claudius at least, first at Rome and afterwards in municipal towns, there had been colleges of *dendrophori* appointed by the government to look after the cult of Cybele. Their name does not imply that they were anything more than "porters"; we have observed that it seems to have been their chief function to secure the sacred pine and *carry* it in religious processions; surely that would not necessitate their being lumber merchants or carpenters. To their religious service they could easily add the civil one suggested above: their name would still be particularly appropriate; for the most natural way for them to transport heavy burdens would be by means of poles carried by several of them from hand to hand or shoulder to shoulder. In this way, then, *dendrophori* also may have become united with the *collegium fabrum et centonariorum*. Apparently some *collegia dendrophorum* sustained an independent existence, no doubt as religious priesthoods. This was probably unusual, however, for the alliance of the three associations was still the rule as late as the fourth century, when Constantine passed an edict directing: "In quibuscumque oppidis dendrophori fuerint, centonariorum atque fabrorum collegiis annectantur."[50] At times the three are referred to as distinct colleges closely related; at others, they seem to be three divisions of one organization. Again, one or even two of the groups may not appear, but this need not necessarily mean that the bodies mentioned did not contain any of the other one or two; for the name may have been adopted from the majority membership.[51]

There is every reason to believe that this collective body took the place in *municipia* of the *cohortes vigilum* at Rome; we may note its omnipresence, its public character, and especially the fact that at Rome,

[49] Dig. 33.7.12.18.
[50] *Cod. Theod.* 14.8.1.
[51] Cp. Waltzing 1.341-346; 4.51-72, 76-80.

where the *vigiles* were originally established, and at Ostia and Puteoli, where according to Suetonius, Claudius later appointed a similar organization,[52] there appears to have been no *collegium fabrum*.[53] At Nimes a *praefectus vigilum* was apparently at the head of a *collegium fabrum*.[54] Now if there was this connection between the two bodies, the municipal Guilds, like the city Watch, would doubtless be entrusted not only with extinguishing fires, but with maintaining law and order. In this case, the policing of the town possibly devolved upon the *dendrophori* particularly; as the bearers of heavy burdens, they would be especially stalwart men; then too, their connection with a religious cult may have assured them greater respect. The *dendrophori*, therefore, may have had some relation to the *hastiferi* found at Vienna, Cologne, and Cassel, to whom Mommsen assigns the duties of municipal police.[55]

This whole theory in regard to the *dendrophori*, which has been set forth in the preceding pages, finds very strong support in certain customs existing in Constantinople at the present day. These are described by H. G. Dwight in an article entitled "Life in Constantinople."[56] In that city, according to Mr. Dwight, the porters have manifold duties and are consequently very important. Divided into guilds and subguilds, they are located in every quarter of the town, and do all the fetching and hauling, carrying by hand or back, or by poles from shoulder to shoulder, any kind of burden from hand luggage to a piano. Some of them also serve as *night watchmen*, and report fires. Most interesting of all is a special guild of *firemen*; they are called *bouloumbajis*, that is, "pumpmen"; for they carry a handpump on a wooden box which has two poles at each end to rest on the men's shoulders. It is their duty also to remove furniture from burning buildings; and as a direct result of this, they consider it their peculiar right to act as furniture movers, even if a family has not been burned out, but wishes, for normal reasons, to make a change of residence. Mr. Dwight particularly notes that some of these customs go back to time immemorial. They may very conceivably, therefore, be a relic that has been handed down from the days

[52] Suet. *Claud.* 25.2.

[53] Liebenam 104; Kühn 28. It is not surprising, however, to find inscriptions of *centonarii* in the Capital; they would be needed of course to supply the army. *Dendrophori* are recorded at all three places, they were apparently religious bodies for the most part. Cp. Waltzing 4.11, 15, 17 f., 77.

[54] Liebenam 104; Waltzing 2.204.

[55] Mommsen, *Die römischen Provinzialmilizien*, Hermes 22(1887).557 f.; cp. Darem.-Saglio 3.1.43; Waltzing 2.152.

[56] Nat. Geog. Mag. 26(1914).521-545, especially 533-539.

of the *dendrophori*, when Constantinople was the Graeco-Roman city of *Byzantium* or *Constantinopolis*.

The *collegium fabrum* with its adjuncts, then, appears to have been a well-equipped and highly organized Department of Public Safety, charged with guarding against fires and upholding the peace. Presumably, the *fabri* made and manipulated the apparatus; the *centonarii* manufactured canvas, piecing it together to make either protective coverings for implements, or cap and tunic uniforms which may have been for the firemen in general, or for the *centonarii* themselves, that they might form a special brigade to fight nearest the flames; the *dendrophori*, a company of porters, attended to all carrying and hauling required by the duties of the department, and probably looked after the policing of the town. At the head of the united or allied bodies was a *praefectus fabrum*, who in this municipal position is to be distinguished from the military aid of the same name that served in the field. However, since there are indications that the colleges to whom the protection of the town was entrusted were under military formation and discipline,[57] the municipal prefect may possibly have been a military officer like the *praefectus vigilum*. Considering his title and his duties, he was doubtless appointed by the government; that is, by the emperor or by one of his functionaries.[58]

There has been a long digression from our discussion of *tignarii*, but since the writer holds a strong conviction against the current belief that these made up the ranks of the *collegia fabrum municipalia*, the foregoing argument in all its detail was deemed necessary in order to disprove the allegation, and to distribute more broadly the prestige previously held by *tignarii* alone. It is true that Kühn's study of inscriptions points to the fact that carpenters and builders as an individual group were probably the most prominent of all *opifices*; the epigraphic evidence which he has collected concerning them tends to show that the greater number were freemen and members of corporations which admitted no slaves and had patrons of high position.[59] *Fabri* in general, however, must have enjoyed the respect and esteem of their fellow men. From their ranks, no doubt, were descended those respectable, even distinguished, Roman families who held the names *Fabricius*

[57] Maué 60 f.; Liebenam 210; Waltzing 2.351-356.
[58] Cp. Maué 72-82.
[59] Kühn 27-31.

or *Faber*,[60] and who doubtless felt as little disgraced by their inherited *nomina* as the bearers of Smith and Smithson at the present day. According to Livy, *fabri* were the first of the industrial orders to be admitted to the army, and they were at once assigned to the first class in the Servian revision.[61] Again, Martial did not disdain to serve *faba fabrorum*[62] on his table but rather enjoyed their unpretentiousness even for a dinner party; to. Persius, too, simple *plebeia prandia* were desirable; he believed only a gourmand would scorn them.[63] Even Juvenal's ironic clause, "His crescunt patrimonia fabris,"[64] casts no slur upon *fabris*; for it is *his* (= *avidis*) that holds the emphatic position in his verse and catches the force of his satire. Finally, we perceive that there was recruited from the orders of *fabri* a municipal organization of great importance, which Waltzing believes to have existed in practically every city of the western Empire.[65] Kühn observes from epigraphic sources that this *collegium* consisted chiefly of freemen, evidently admitted no slaves, and was apparently held in the greatest esteem, as is evidenced by the high rank of its patrons.[66] The existence of such a corporation, offering *fabri* no mean share in civic life, shows that trust and responsibility could be placed in them; and membership in the college must have been considered a worthy honor.

XXVI

Tonsores

The Romans applied the term *tonsor* to both shearers[1] and barbers, so that Varro found it not inconsistent to discuss the latter in his treatise on farming, in the midst of an exposition on shearing sheep! He states that the tonsorial art was not practiced in Italy until 300 B. C., when P. Ticinius Menas introduced barbers from Sicily; in support of his

[60] In such inscriptions as CIL. 11.2067, C.PETRONIVS / SEX.F.FABER, Kornemann 6.2.1892 interprets *faber* as designating a freeman member of a *colleg. fabr.* Mommsen, CIL. 5, p. 1199, and Kühn 24 f. more naturally consider it a cognomen which implies no necessary indication of occupation. But after all it no doubt registers the fact that the first members of the family to bear the name were *fabri*.
[61] Liv. 1.43.3; Cic. *Rep.* 2.39. Dionys. Hal. 7.59 consigns them to the second rank.
[62] Mart. 10.48.16.
[63] Pers. 5.17 f., this may contain an allegory as well as a metaphor; cp. 8.111-114.
[64] Juv. 14.116.
[65] Waltzing 2.199 f.
[66] Kühn 34-36.
[1] Mart. 7.95.12 f.; 8.50(51).11; cp. 11.84.17 f.

assertion, he bids us observe that the statues of the ancients usually have long hair and beards.[2] The elder Pliny furnishes this same information in a context even more curious than that of his authority. After maintaining that at a very early period there was a tacit consent among all nations to adopt the letters used by the Ionians, he continues with the declaration that the next point upon which humankind agreed was the employment of barbers, but that the Romans were somewhat slow to join the *gentium consensus!* He adds several interesting items to the effect that Scipio Africanus was the first Roman to shave daily, and that Divus Augustus was always smooth shaven.[3]

Throughout the first century of the Empire,[4] barbers were apparently an essential to every community however small and retired,[5] for as a common custom, except in token of mourning or calamity,[6] none but philosophers wore beards, and only eccentric poets allowed their hair to grow long.[7] Much attention was paid to the care of the head; for an uneven tonsure was a cause of riducule,[8] and hapless indeed was he whose baldness made a barber's services unnecessary! According to Martial, various tricks were tried to conceal the defect, even to painting hair on the bare scalp! To quote his own words:

> Mentiris fictos unguento, Phoebe, capillos
> et tegitur pictis sordida calva comis.
> tonsorem capiti non est adhibere necesse:
> radere te melius spongea, Phoebe, potest.[9]

Domitian was so sensitive about his baldness, says his biographer Suetonius, that he considered it a personal insult for anyone else to be twitted about this defect in jest or in earnest. He published and dedicated to a friend a book "On the Care of the Hair."[10]

There was doubtless a barber and manicure—the *tonsor* served as both[11]—in practically every household that possessed slaves.[12] Martial

[2] Varro *Rust.* 2.11.10.
[3] Plin. *Nat.* 7.210 f.; cp. Suet. *Aug.* 79.1.
[4] Hadrian, we are told, brought beards again into vogue, cp. Dio 68.15.
[5] Mart. 2.48.2.
[6] Sen. *Dial.* 11.17.5; Mart. 2.36.3; 2.74.3; Suet. *Aug.* 23.2; Dio 48.34.
[7] Hor. *Sat.* 2.3.16 f.; *Epist.* 2.3.299-301; Mart. 11.84.7.
[8] Hor. *Epist.* 1.1.94 f.; cp. *Sat.* 1.3.31.
[9] Mart. 6.57; cp. 3.74; Suet. *Otho* 12.1.
[10] Suet. *Dom.* 18.2.
[11] Plaut. *Aul.* 312; Mart. 3.74.3; 14.36; cp. Hor. *Epist.* 1.7.50 f.
[12] Cp. Mart. 6.52; 8.52; 11.58.

suggests tonsorial implements as a suitable gift at the Saturnalia, and offers the following inscription to be appended:

> Tondendis haec arma tibi sunt apta capillis;
> unguibus hic longis utilis, illa genis.[13]

Specimens of these instruments[14] that have been unearthed, especially razors, present such a formidable appearance, that little wonder need be expressed at their being termed *arma*[15] and *barbara tela*;[16] and yet in all fairness be it added that the same words, with their clever pun, seem even more applicable to similar implements of the present day.

The epigrammatist lets his wit flow unrestrained upon the wielders of these weapons. Although he duly recognizes and praises the skilful slave,[17] he more often finds cause for ridicule in the barber with the brigand's nature,[18] the reckless *tonsor* to whom any torture is preferable,[19] and the dawdler whom he sarcastically names "Nimble."[20]

Besides the barbers who were slaves in personal service,[21] there were others who conducted an independent business in shops, *tonstrinae*.[22] They must have been fairly numerous to admit the possible truth of Horace's statement that the restless tendency of his time asserted itself even in the *pauperes*, forcing them to change constantly their *cenacula, lectos, balnea, tonsores*.[23] Business was not always confined indoors; for according to Martial, barbers blindly plied their razors amid the crowds of the open street and sidewalk, until Domitian's law restrained shopkeepers behind their thresholds.[24] It was seldom, we may suppose, that *tonstrinae* were quiet and deserted, unless it

---

[13] *Id.* 14.36.

[14] Hor. *Epist.* 1.7.51; Petron. 94, 103, 108; Plin. *Nat.* 7.21; Mart. 7.61.7; 7.95.12; 8.52.7; 9.76.5; 11.58.5; 11.84.3 mention knives, *cultri, cultelli*; shears, *forfices*; razors, *novaculae*; mirrors, *specula*; razor-cases, *thecae*; tweezers, *volsellae*. Cp. Nicolson 51; Blümner, Müller's *Handbuch* 4.2.2. 267-269.

[15] Mart. 14.36.1; cp. Petron. 108.

[16] Mart. 11.84.12.

[17] Mart. 6.52; 8.52, cp. CIL. 6.11931.

[18] Mart. 11.58.5-10; cp. Petron. 94.

[19] Mart. 11.84.

[20] *I.e.*, Eutrapelus <εὐτράπελος, Mart. 7.83.

[21] Eumolpus, in Petron. 94, 103, 108, had a barber who was a hired servant, *mercennarius*.

[22] Plaut. *Amph.* 1013; *Epid.* 198; Ter. *Phorm.* 89; Petron. 64. Hor. *Epist.* 1.7.50 speaks poetically of a barber's booth as an *umbra tonsoris*.

[23] Hor. *Epist.* 1.1.91 f.

[24] Mart. 7.61.7, 9. See p. 61.

was in the early part of the afternoon.[25] Their patrons, however, were not always on serious purpose bent; for "tonsorial parlors" in both Athens and Rome, like their modern counterparts in most communities, were popular places to meet one's friends and gossip.[26] So it is that Horace with no little humor remarks:

> Proscripti Regis Rupili pus atque venenum
> hybrida quo pacto sit Persius ultus, opinor
> omnibus et lippis notum et tonsoribus esse.[27]

And one of Trimalchio's friends boasts that in his racing days, before he had the gout, he could sing, dance, recite, and act the clown in general in imitation of the chatter in a barber shop.[28] *Tonstrinae* must of course have been located in all parts of the town. A locality in the vicinity of the Temple of Flora, on or near the Quirinal, seems to have been known as *ad tonsores*.[29] Horace mentions a barber's booth in or near the Forum, which the lawyer Philippus passed on his way from the law courts to his home on the Carinae.[30] Interesting in connection with this is the following epigram from Martial:

> Tonstrix Suburae faucibus sedet primis,
> cruenta pendent qua flagella tortorum
> Argique letum multus obsidet sutor.
> sed ista tonstrix, Ammiane, non tondet,
> non tondet, inquam. Quid igitur facit? Radit.[31]

Since the Argiletum and Subura were probably the main thoroughfares of communication between the Esquiline and the central part of the city,[32] it may be that both Horace and Martial had in mind the same general district. The allusion to a woman barber here is noticeable;[33] if she is a true example of her kind, we must infer that they were of low character and enjoyed little respect.[34]

[25] Cp. Hor. *Epist.* 1.7.47-50.
[26] Lysias *Orat.* 24.20; Demos. *vs. Phorm.* 13; Plaut. *Amph.* 1013; *Epid.* 198; Ter. *Phorm.* 89 ff. See p. 73.
[27] Hor. *Sat.* 1.7.1-3.
[28] Petron. 64.
[29] CIL. 15.7172; Platner 486, 489.
[30] Hor. *Epist.* 1.7.46-51.
[31] Mart. 2.17. See pp. 59 f.
[32] Gilbert 1.162-164; Jordan 1.3.262-265; Platner 40,444,446.
[33] Cp. Plaut. *Truc.* 405, 772, 856. Abbott, *Society and Politics*, 98, on "Roman Women in the Trades and Professions," says: "Women of the lower classes entered freely into the medical profession and the trades."
[34] On the ill repute of women in the Subura, see pp. 65 f.

The question arises as to whether *tonsores* in general were as disreputable as the *tonstrix* of Martial's verses. If we turn to the satirists for reply, we find references to several barbers who prospered very much in a material way, but apparently not at all socially. Horace, for instance, speaks of a certain Alfenus who, having laid aside his instruments and shut up shop, turned lawyer perhaps, or usurer, at all events, one to be characterized by the adjective *vafer*.[35] Some may note an added thrust in the poet's declaration that once a barber meant always a barber, but this would be rather overstepping the mark; for the special point that he wishes to make is that a man may be actively engaged in one pursuit yet remain potentially the master of another.[36] Horace also takes passing note of Licinus,[37] who was presumably a popular barber of his day. Because of similarity of name, possibly, and of analogy to other *tonsores* recorded in the verses of Martial and Juvenal,[38] Acron identified this Licinus with the freedman of Caesar and Augustus,[39] who was appointed by the latter as procurator of Gaul, amassed a great fortune comparable to that of Crassus, and left a marble tomb to recall his memory to many generations of noble Romans.[40] The identification is highly improbable; for we should expect to find a hint of this early occupation in references which are clearly to the procurator Licinus, as we do in the case of Nero's freedman, *Vatinius, sutor Beneventinus*.[41]

Juvenal goes so far as to exclaim that he was forced to write satire, when in addition to other unendurable conditions, he observed one man in possession of countless villas and vying in wealth with all the nobility, although he had once been one "quo tondente gravis iuveni mihi barba sonabat." This scornful verse occurs verbatim in two Satires.[42] Some editors[43] suggest, although it seems a matter of very broad conjecture, that Juvenal is referring to the same man whose character Martial tears

[35] Hor. *Sat.* 1.3.130-132.
[36] This passage may refer to a shoemaker. The original reading of V for v.132 is *tonsor*; other MSS. have *sutor* or a corruption. Modern editors in the main have adopted the first; the scholiasts, however, read *sutor* and treated the lines as a specific allusion to Alfenus Varus, a shoemaker of Cremona, who became a senator at Rome; see p. 58, n. 27.
[37] Hor. *Epist.* 2.3.301.
[38] See nn. 42, 45.
[39] *Schol.* on Hor. *l.c*
[40] Pers. 2.36; Sen. *Epist.* 120.19; Mart. 8.3.6; Juv. 1.108 f.; 14.305-308; Dio 54.21.
[41] See pp. 58 f.
[42] Juv. 1.25; 10.226.
[43] Cp. Duff, on Juv. 1.24; Bridge and Lake, on Mart. 7.64.4.

to shreds under the name of Cinnamus.⁴⁴ This fellow, formerly a slave, well known as a barber all over town, had actually become a knight. He had probably acquired much of his fortune by acting as one of the Emperor's *delatores*; for he later fled to Sicily to avoid trial at Rome. Martial directs a stinging epigram against him; he recalls his past, assures him that he can hope to find no joy in the unhappy leisure of a fugitive, but reminds him that since he is without education and is too far away from the Capital to rely upon his knavish flattery, he is fit to become nothing but a barber again.⁴⁵ He is apparently the target also of the two-edged shaft of 6.17, where the epigrammatist comments adroitly upon his attempt to obliterate all traces of his former station, proclaiming:

> Cinnam, Cinname, te iubes vocari,
> non est hic, rogo, Cinna, barbarismus?
> tu si Furius ante dictus esses,
> Fur ista ratione dicereris.⁴⁶

Although it may be argued that the foregoing satire is aimed primarily at avarice and deception, or is possibly intended for specific individuals, still we must admit that it appears to attack the occupation as well as the man. It would seem, therefore, that *tonsores* had a somewhat toilsome and precarious struggle for social existence. We recall at once that barbers were not to be found at Rome in the infancy of the state, when as Wezel claims, very many of the artisans and tradesmen were freeborn;⁴⁷ they had, therefore, no guild of long standing to grant them a certain degree of prestige and special privilege.⁴⁸ On the contrary, their trade had been introduced as a slave's employment and had doubtless continued to draw largely from this source. It is

---

⁴⁴ Giese 11 considers this a fictitious appellative.

⁴⁵ Mart. 7.64.

⁴⁶ Cp. Mart. 6.64.26: "Stigmata nec vafra delebit Cinnamus arte." This is explained by Friedländer in his ed. as a reference to a physician. It seems quite consistent, however, to refer it to the *tonsor* of 7.64 and 6.17. In 6.17, quoted above, note the excellent, triple pun in *barbarismus*: "a barbarism," "a barbarous act," "a barber's act" (*i. e.* cutting out undesirable features). Nicholson 43 maintains that barbers cut corns and removed warts and other corporeal disfigurements. The art of *tonsores* in Rome was doubtless as expansive as it was formerly in England, for instance, where, according to the New English Dictionary 1.666, "the barber was also a regular practitioner in surgery and dentistry"; in 1461 a company of Barber-Surgeons was incorporated by Edward IV; the barber's pole is a reminiscence of his surgical activities, cp. Encyclopaedia Britannica.

⁴⁷ Cp. Wezel 31 f.

⁴⁸ Cp. Typaldo-Bassia 52-58, "De l'ouvrier libre incorporé."

altogether probable that independent barbers were almost exclusively *liberti* at best,[49] and as there must have been every type from the *inaequalis tonsor*, who was to be eschewed for his uneven work and temperament,[50] to the Master Barber, *tonsor magister*, who might be sought upon special occasions,[51] many of them were doubtless subject to the same ridicule which Martial directs against barber slaves.[52] It may be, therefore, that the calling of *tonsores* was hampered by its servile origin and associations, and that for this reason the efforts of barbers for social recognition met with greater resistance. Some, however, evidently attained a goal sufficient to yield a fair competence and bring contentment and satisfaction; one of Petronius's *parvenus*, who had acquired success himself and entertained aspirations for his son, determined to have him take up the vocation of barber, auctioneer, or advocate at least, if he should shrink from the profession of jursiconsult;[53] and *tonsores* at Pompeii took an active interest in municipal elections.[54]

[49] It happens that Kühn's register of *Tonsores*, 66, contains no inscriptions for *ingenui*, but this may be mere chance, for none are recorded for slaves either.

[50] Hor. *Epist.* 1.1.94.

[51] Juv. 6.26. Friedländer, in his ed. of Juv., takes this phrase as evidence that there was a school where the art of hair dressing was taught just as there was one for meat carving (Juv. 11.136-144). This is very possible; for Petron. 94, recounting an attempt at suicide on the part of Giton and Encolpius, says that the razor which they seized for the purpose, proved to be one that was not sharp, but had been especially blunted in order to give young pupils (*pueris discentibus*) a barber's boldness, and had been enclosed in a case. It seems quite as natural, however, and a little simpler to interpret *magister* in our passage as one who had thoroughly "mastered" his art and was therefore considered "head" or "chief" of all the *tonsores* in town or among the household slaves. Perhaps the term was suggested because of its regular technical use for the "head" of an industrial corporation or a body of specialized slaves. Cp. Waltzing 1.388-405; 4.341-349.

[52] See p. 89. Mart. 7.83 and 11.84 probably refer toi ndependent barbers.

[53] Petron. 46.

[54] CIL 4.743.

CONCLUSION

As a result of the foregoing investigation, we are led to believe that Rome's industrial population played a significant part in the life of the early Empire and received no inconsiderable recognition. It is true that the favorite occupations were still agriculture, law, and war.[1] The satirists naturally pleaded also the cause of literature;[2] and Martial, whose indolent tendencies inspired in him an admiration for Saturn's reign, "sub quo pigra quies nec labor ullus erat,"[3] numbered a client's attentions among the most honorable and desirable means of livelihood.[4] But the growing power of wealth and the commercializing of the old aristocratic pursuits[5] were extending their influence broadcast, so that the Ciceronian attitude toward paid labor[6] could no longer be rigidly sustained. Even the Republic had seen the domination of the knights, Rome's financiers and business men; but the Empire was characterized by the rise of a much lower order, when wealthy freedmen mounted to high positions of influence.[7] With the manifold opportunities open to these *liberti* in the government,[8] and the lessening competition of slave labor, due to extended periods of peace which diminished the supply of slaves but increased the call for workers,[9] the incentive, indeed the necessity, for freemen to enter the industrial ranks was very great.

And certainly the outlook for trade and industry had never been brighter. The craze for building that caught the Roman world, from the emperors down to men of private station,[10] must have furnished steady employment to hundreds of *fabri* whose efforts were directed to construction or adornment. Extravagance and lavish expenditure,[11]

---

[1] Hor. *Sat.* 1.1.4-12; *Epist.* 2.3.314 f.; Mart. 2.64.1; 3.33.3-6; 10.15(14).6; 12.16.1; 14.34; Juv. 8.47-52, 79, 87-89; 14.70-72; 16.1 f.
[2] Hor. *Carm.* 1.1.29-36; Mart. 3.38.7-10; Juv. 7.
[3] Mart. 12.62.2.
[4] *Id.* 3.38.11.
[5] Hor. *Sat.* 1.1.28-35; Petron. 46, 83; Juv. 14.189-198. Cp. Cauer 698.
[6] Cp. Cic. *Off.* 1.150; Miller 12-15.
[7] Petron. 38: "Liberti scelerati . . . omnia ad se fecerunt"; cp. Cunningham 173.
[8] Friedländer-Magnus 1.33-56.
[9] Duruy 6.289; Typaldo-Bassia 58-59; Cauer 699; Tucker 8-15.
[10] See p. 79, n. 19.
[11] Friedländer-Freese 2.131-230; Davis 152-193.

too, proved the harvest of many, no doubt, besides the contractor and builder. The importer of foreign materials and commodities would reap rich profits; cabinet makers and workers in metal would have much to do to meet the demand for furniture, decorative ornaments, gold and silver plate, and jewelry.[12] Juvenal even suggests the dismal fear that the constant call upon the *ferrarius* for iron chains and shackles from every forge and anvil would cause the supply of ploughshares, mattocks, and hoes to fail![13] Again, there must have been urgent need of fulling and dying establishments to dye the cloth and to attend to the scouring, cleaning, pleating, and pressing required for the togas, syntheses, and lacernae of many people who would find it too inconvenient or expensive to employ their own *fullones* and *infectores*.[14] For shopkeepers the opportunities seem to have been especially good: the desire for luxuries of every description brought fine shops to the Sacra Via, the Saepta,[15] and the porticoes;[16] and the spendthrift prodigals, who wasted a patrimony, or two, in riotous living, were a source of great blessing to the fish and poultry man, the fruitier, the delicatessen dealer, and all the tradesmen of the Vicus Tuscus and the Velabrum.[17] Then too, a city so full of apartment houses that even in Cicero's time it could be described as *cenaculis sublatam atque suspensam*[18] must have contained many persons of moderate means who could afford only two or three slaves, or none, and who would therefore give ample patronage to the ordinary butcher, baker, and dealer in general utilities.

We may concede, then, that under the early Roman emperors industry prospered, and with it those who were engaged therein. So it was that another factor was added to help overcome the aversion to entering the trades and professions. It was apparently not uncommon for even the humblest tradesmen and craftsmen to acquire a fortune and retire from active business, or change their former pursuit for a more leisurely occupation. Martial, as we have noted, admitted that there was an argument for vocational training versus a liberal education in the success of the *praeco*[19] and the *sutor*.[20] Had not a shoemaker at Bononia ex-

[12] Cp. Petron. 32 f., 19 f., *et passim*; Mart. 3.62; 8.6; 14.89 ff.
[13] Juv. 3.309-311.
[14] Cp. Marquardt 2.504-516; 527-530. See p. 21.
[15] See pp. 69 f.
[16] Mart. 10.87.9 f.; Juv. 6.153-157. See pp. 33, n. 29; 38.
[17] Hor. *Sat.* 2.3.226-238. See pp. 66 ff.
[18] Cic. *Leg. Agr.* 2.96.
[19] Mart. 5.56. See p. 52.
[20] Mart. 9.73. See p. 58.

hibited gladiatorial shows,[21] and Vatinius, the influential favorite of Nero, been a *sutor* of Beneventum?[22] And was it not possible for an auctioneer to win a wife though he had as his rivals two praetors, four tribunes, seven advocates, and ten poets?[23] Trimalchio's friend Echion, with all his aspirations for his sons, considered that one of them had become steeped quite enough in literature, and determined to make of him a barber, an auctioneer, or "at least" an advocate.[24] Juvenal, too, vouches for the possibilities open to the *tonsor*; for he had seen the one to whom he had gone when a young man become the possessor of many villas;[25] and Martial records a barber who had come into a knight's fortune.[26] We have seen also that a *fullo* exhibited gladiatorial shows at Mutina;[27] and that Crispinus, the favorite of Domitian, was made *princeps equitum*, although according to report, he had once been a dealer in salt meat and fish.[28]

Indeed, as Petronius affirms, men rose from nothing: one who, but a short time ago, used to carry wood on his back, soon counted his eight hundred thousand sesterces;[29] another, starting with a copper, left a solid hundred thousand sesterces and all in cash.[30] Trimalchio's progress, as portrayed by his chronicler, was thoroughly miraculous. He began as a slave from Asia; he became his master's accountant and steward, and after his emancipation, entered into mercantile pursuits. Under the sure guidance of Mercury and Minerva, he climbed to a lofty pinnacle of wealth and influence. As a *sevir Augusti*, one of the six officials who were appointed annually in municipal towns to be responsible for the cult of the emperor, he was privileged to wear a gold ring and the *toga praetexta*, to have two lictors, and to sit on a throne.[31]

---

[21] Mart. 3.59; cp. 3.16, 99. See pp. 57 f.
[22] Mart. 10.3.4; 14.96; Juv. 5.46 and *schol.*; Tac. *Ann.* 15.34. See pp. 58 f.
[23] Mart. 6.8. See p. 51.
[24] Petron. 46. See pp. 51, 81 f., 93.
[25] Juv. 1.24 f.; 10.225 f. See p. 91.
[26] Mart. 7.64; cp. Hor. *Sat.* 1.3.130-132. See pp. 91 f.
[27] Mart. 3.59. See p. 21.
[28] Juv. 1.26-29; 4.28-33, 108 f. See p. 27.
[29] Petron. 38. See p. 25.
[30] Petron. 43.
[31] *Id.* 29 f., 65, 71; cp. Darem.–Saglio, Ruggiero's *Dizionario Epigrafico*, and Pauly-Wissowa under *Augustales*, and L. R. Taylor, *Augustales, Seviri Augustales, Seviri*, T. A. P. A. 45(1914).231 ff.

He owned so many slaves that scarcely a tenth part of them knew him by sight. His palatial residence at Cumae(?) had four dining-rooms, twenty bed-rooms, two marble colonades, and every convenience; and wherever the kite flew, there were other estates in his possession. In short, he lived like a prince in the midst of affluent luxury, so fabulously rich that he himself did not know the amount of his wealth; yet as he estimated it for his epitaph, he could count on leaving thirty million sesterces at his demise.[32]

But strange rumors were in circulation about even the emperors themselves, and although these may have been mere idle tales, they must at least have been probable or they would have had no point. It was said that the paternal great-grandfather of Augustus was a ropemaker, and that his grandfather was a broker; while, on his mother's side, his great-grandfather had been at one time a perfumer, later a baker.[33] One account had it that the great-great-grandfather of Vitellius had been not merely a *sutor*, but a *sutor veteramentarius*; and that his great-grandfather had married the daughter of a baker; yet his grandfather became an *eques Romanus*, his father a senator, and he himself the ruler of the Roman world![34] Vespasian's relations with business were closer still: he was actually nicknamed "Muleteer," because he had once taken up dealing in mules in order to pay off his debts and "maintain his dignity;" and even when emperor, he openly engaged in trade.[35]

Now although Suetonius's "gossip" is more or less colorless, it must be admitted that the satirists cite their examples with no little irony. But sarcasm does not alter the facts; we must recall the old adage, *facta non verba*. Just as Tacitus gloomily condemned *sordidae merces*[36] two thousand years ago, so our theorists and moralists to-day discourse on "filthy lucre" and "tainted money" and an English editor writes of Trimalchio's friends at Cumae (?), that men became "millionaires with American rapidity;"[37] yet not only business men, but tradesmen and craftsmen continue to prosper, and to hold positions of prominence in their communities politically, economically and socially; and the families of American millionaires are presented at the English court. We have every reason to believe that similar condi-

---

[32] Petron. 29, 32 f., 37 f., 53, 71, 75 *fin*. ff.
[33] Suet. *Aug.* 2.3; 4.2. See p. 43.
[34] Suet. *Vitel.* 2.1 ff.
[35] *Id. Vesp.* 4.3; 16.1.
[36] Tac. *Ann.* 4.13.4; cp. 4.62.2.
[37] Heseltine, ed. of Petron., Introd. xi.

tions existed at Rome in the first centuries of the Empire, when the great leveling influence of imperialism was potent to raise the humble and weaken the haughty.[38] We are too prone to jump at conclusions from the exaggerations of the satirists,[39] forgetting that their criticism is of *excesses* along various lines. It is not fair to assume that all who met with phenomenal success in business became ostentatious and vulgar like Trimalchio and his friends; but there would be no cause for satire in the case of the *un*pretentious, so why write of them? And after all, Petronius's effusion, for all its ridicule, may show just how possible it was for an upstart, even a former slave, to gain apparent, if not actual, recognition in society. If the author was, as is the commonly accepted view,[40] the dashing Gaius of Nero's court, and if his information was from first hand knowledge, he, the *elegantiae arbiter*,[41] must have associated intimately with such people as he describes; if, on the other hand, he was writing merely as an observer from superior heights, his words must be discounted all the more. At best his novel is but a travesty.

Upon the condition of that vast number of workers who did not aspire to higher position, but remained in the industrial ranks at home or abroad to supply the daily wants of the Roman Capital and muncipalities, we feel that we need waste little commiseration. Many freemen no longer hesitated to make money by trade;[42] and the most conservative must gradually have been led to see, as Juvenal was, that a livelihood earned through honest business was more befitting a freeborn man than that gained through obsequious sycophancy,[43] in which even members of the higher orders indulged.[44] Inevitably the humblest pursuits were in the hands of the lower classes, but people of standing engaged in industries organized on a large scale; the brick business, for instance, as is seen from inscriptions, was in great measure controlled by women of leading families.[45]

Of course the working classes still contained a large number of freedmen, but it is altogether probable that the line of distinction between *liberti* and *ingenui*, and their families, was not so closely marked

---

[38] Cp. Friedländer-Magnus 1.33.
[39] Cp. Duruy 6.302-308.
[40] Teuffel-Schwabe-Warr 2.§ 305.4; Schanz 129 f.
[41] Tac. *Ann.* 16.18.
[42] See lists in Kühn *passim*; cp. Juv. 14.201 ff.
[43] *Id.* 7.3-16; cp. Petron. 116. See p. 39, n. 65.
[44] Mart. 12.29(26); Juv. 3.126-130.
[45] See p. 20.

as we sometimes allow ourselves to imagine, especially since both ranks were often represented in a single household. The story of Horace, the freedman's son, attending school like any senator's child,[46] is familiar to all. Then there is the amusing incident of Encolpius's discomfiture at Trimalchio's dinner, when, thinking that the praetor had arrived, he would have sprung deferentially to his feet, had he not been told that the newcomer was only a friend of his host's, Habinnas, a pompous mason![47] And Martial makes frequent allusion to the trials of the ushers at the theatre in reserving the first fourteen rows of seats for the *equites*; for freedmen, and even slaves, were able by some trick of dress or manner to pass off as knights and get in unnoticed.[48] One of Trimalchio's guests definitely asserts: "For forty years I was a slave, but no one knew whether I was bond or free." Several other remarks of this man are highly interesting; from them we infer that young provincials sometimes became slaves voluntarily because of the prospect of future manumission and citizenship. "Of my own accord," says our informant, "I gave myself into slavery, preferring to be a Roman citizen rather than a provincial tax-payer. Now I hope my life is such that nobody can make sport of me. I am a man among men, I walk about with my head uncovered; I owe no one a copper . . . I was a boy with long curls when I came to this town . . . but I made every effort to prove satisfactory to my master . . . and in spite of opposition in the house, I stemmed the flood to the finish. These are true victories; for to be born free is as easy as saying 'Come here' ".[49] Apparently freedmen were not infrequently the guests of freemen; for there seems to have been a special place at the table, which was assigned to them. Petronius for instance, speaks of someone who was "reclining in the freedman's place" ("qui libertini loco iacet").[50] Furthermore, since we are not without encomiums on admired and respected freedmen, examples of which are Horace's tribute to his father,[51] and Persius's noble lines on his friend and teacher Cornutus,[52] we may justly suppose

[46] Hor. *Sat.* 1.6.71-82.

[47] Petron. 65. Habinnas was preceded by a lictor, because he was a *sevir Augustalis*; see p. 96.

[48] Mart. 5.8, 14, 23, 25, 35, 41; 6.9.

[49] Petron. 57.

[50] *Id.* 38. Cp. Miller 2-7, especially 6.

[51] Hor. *Sat.* 1.6.65-99.

[52] Pers. 5.30-51.

that there were innumerable other *liberti* whose merits, though not immortalized, were none the less appreciated.

Naturally, various types of individuals were included in the industrial orders then as now, the indigent and dishonest, the energetic and scrupulous. The former would be eschewed, and having little respect for themselves, could scarcely hope to receive it from others. The latter, typified, perhaps, by the *praeco* of Horace's well-known Epistle, no doubt lived simply and contentedly, often in homes of their own, with congenial friends, and with sufficient opportunity not only for industry but for recreation in the Campus or at the games.[53] Certain tradesmen and craftsmen, primarily perhaps because of the long association of their occupations with slaves, appear to have met with little esteem; for example, *caupones*, *coci*, *institores*, *textores*, *mangones*, and to a degree, *tonsores*. Others, however, especially those who enjoyed the privileges of corporations that dated from early times, undeniably received due consideration, as may be judged in many cases from the prominence of their patrons.[54] It is interesting to note that the hostile legislation against *collegia* at the close of the Republic and at the beginning of the Empire[55] spared the guilds ascribed to Numa and similar early associations, whose very antiquity seems to have conferred upon them a certain prestige. Augustus, perceiving with his characteristic insight, that labor organizations supplied a human need and were useful to the state as well, established a policy, not of annihilation, but of state supervision; he even granted special privileges to *collegia* which were incorporated for the public utility. His lead was followed quite uniformly by subsequent emperors of the first century after Christ. Epigraphic evidence shows the enormous scope of labor guilds at this period,[56] while Pompeian wall *graffiti* bear witness to the influence that they exerted upon municipal elections.[57] The fact that in the second century the regulating hand of the emperors began to tighten upon them implies that they had been gaining too much power.[58] All working men of course were not incorporated; but as Typaldo-Bassia remarks, in

[53] Hor. *Epist.* 1.7.55 ff. Cp. Tucker 253-257. See pp. 52 f.
[54] Cp. Waltzing 1.427.
[55] Ascon. *Corn.* 75.67 (Clark); Suet. *Iul.* 42.3; *Aug.* 32.1.
[56] Cp. Waltzing 4.1-128 for a collection of inscriptions of industrial *collegia*; also Dessau 2.2.737-760, and Kühn's lists.
[57] Cp. Abbott, *Society and Politics*, 3-21, on "Municipal Politics in Pompeii."
[58] Industrial *collegia* are discussed in brief compass by Typaldo-Bassia 52 ff. and by Abbot, *Common People*, 205-234: "Reflections on some Labor Corporations."

view of the advantages to be derived, it was probably exceptional, at least for free *opifices*, not to be members of *collegia*.[59] In addition to an eagerness for forming guilds, the industrial classes made further manifestation of the pride which they felt in their work, by idealizing it for mural decoration, as seen in the Cupid and Psyche pictures in the house of the Vettii at Pompeii,[60] and by having the implements of their business carved on their tombstones as a lasting memorial.[61]

We may readily conclude, therefore, that the first and second centuries of our era saw a revival of industrial life in the Roman world in both town and country.[62] More freemen were probably engaged in the trades and crafts than ever before, and it was perfectly possible for shrewd and very ambitious *opifices* to acquire a fortune, retire from business, and vie with men of higher birth; furthermore, the majority of those who continued to fill the ranks of steady toilers in homely pursuits, apparently felt an honest pride in their work, maintained flourishing corporations, took an active interest in public affairs, and lived, for the most part, happily and contentedly. Humble, but not degraded, they realized in their totality the force of Horace's words:[63]

> Metire se quemque suo modulo ac pede verum est.

[59] Typaldo-Bassia 19.
[60] Mau-Kelsey 331-337. Cp. Petron. 29. See p. 73, n. 123.
[61] Cagnat, Darem.-Saglio 1738; Duruy 6.291 f.; Tucker 254, fig. 75 Cp. Petron. 71. See p. 44.
[62] Cp. Cunningham 178.
[63] Hor. *Epist.* 1.7.98.

RENEWALS 458-4574

DATE DUE